Worksheets

For Classroom or Lab Practice

Cindy Trimble & Associates

with contributions from **Jeanette Shotwell**
Central Texas College

Beginning &
Intermediate Algebra

FOURTH EDITION

Elayn Martin-Gay

PEARSON

Prentice
Hall

Upper Saddle River, NJ 07458

Editorial Director, Mathematics: Christine Hoag
Editor-in-Chief: Paul Murphy
Sponsoring Editor: Mary Beckwith
Assistant Editor: Georgina Brown
Senior Managing Editor: Linda Mihatov Behrens
Associate Managing Editor: Bayani Mendoza de Leon
Project Manager, Production: Robert Merenoff
Art Director: Heather Scott
Supplement Cover Manager: Paul Gourhan
Supplement Cover Designer: Victoria Colotta
Operations Specialist: Ilene Kahn
Senior Operations Supervisor: Diane Peirano
Mgr. Visual Research and Permissions: Karen Sanatar

© 2009 Pearson Education, Inc.
Pearson Prentice Hall
Pearson Education, Inc.
Upper Saddle River, NJ 07458

The author and publisher of this book have used their best efforts in preparing this book. These efforts include the development, research, and testing of the theories and programs to determine their effectiveness. The author and publisher make no warranty of any kind, expressed or implied, with regard to these programs or the documentation contained in this book. The author and publisher shall not be liable in any event for incidental or consequential damages in connection with, or arising out of, the furnishing, performance, or use of these programs.

Printed in the United States of America
10 9 8 7 6 5 4 3 2 1

ISBN-13:   978-0-13-603087-4   Standalone

ISBN-10:   0-13-603087-4   Standalone

ISBN-13:   978-0-13-603088-1   Component

ISBN-10:   0-13-603088-2   Component

Pearson Education Ltd., London
Pearson Education Singapore, Pte. Ltd.
Pearson Education Canada, Inc.
Pearson Education—Japan
Pearson Education Australia PTY, Limited
Pearson Education North Asia, Ltd., Hong Kong
Pearson Educación de Mexico, S.A. de C.V.
Pearson Education Malaysia, Pte. Ltd.
Pearson Education Upper Saddle River, New Jersey

Contents

Name: Date:

Instructor: Section:

Chapter 1 Review of Real Numbers
Section 1.2 Symbols and Sets of Numbers

Learning Objectives
1. Use a number line to order numbers.
2. Translate sentences into mathematical statements.
3. Identify natural numbers, whole numbers, integers, rational numbers, irrational numbers, and real numbers.
4. Find the absolute value of a real number.

Vocabulary.
Use the choices to complete each statement.

Absolute value	**Natural numbers**	**Whole numbers**
Inequality	**Rational numbers**	**Zero**
Integers		

1. _____ is neither positive nor negative.

2. The _____ are the entire set of natural numbers and 0.

3. _____ are numbers that cannot be expressed as a quotient of two numbers.

4. The distance a real number c, is from zero is called the _____. It is denoted by

    _____ .

5. The symbols $\neq$, $\leq$, and $>$ are called _____ symbols.

6. The _____ are all the numbers in the set {1, 2, 3, 4,...}

7. The _____ are the set of numbers that can be written as a quotient of two numbers.

8. The set of _____ are all the numbers that can be represented on a number line.

9. The _____ are the set of all the positive and negative whole numbers.

Objective 1
Insert <, >, or = in the appropriate space to make the statement true.

10. 15 ___ 12

11. 7 _____ -10

12. -3 _____ -2

13. 1.24 _____ -2.24

10._____

11._____

12._____

13._____

Are the following statements true or false?

14. $11 \leq 11$

15. $3(2+4) < 2(3+4)$

16. $24 > 34$

17. $-1.78 \leq -1.68$

14._____

15._____

16._____

17._____

Objective 2
Write each sentence as a mathematical statement.

18. Three is less than or equal to six.

18. _____

19. Negative eight is greater than negative fifteen.

19. _____

20. Seven is not equal to negative seven.

20. _____

21. Five is greater than or equal to four.

21. _____

Objective 3
Tell which set(s) of numbers each of the following belong to: natural numbers, whole numbers, integers, rational numbers, irrational numbers, and real numbers.

22. 5

22. _____

23. -8

23. _____

24. 1.7

24. _____

25. $\dfrac{2}{3}$

25. _____

26. $\sqrt{14}$

26. _____

Use integers to represent the values in each statement.

27. Since I started my diet, last month, I have lost 6 pounds.

27. _____

28. When balancing her checkbook, Jane found that she was overdrawn by $20.

28. _____

29. The weather man on channel six news promised that there will be an increase of 15 degrees over the next few days.

29. _____

30. During a Chicago Bears game, the quarterback threw a pass, to help the team gain 45 yards.

30. _____

Objective 4
Find the absolute values of each number.

31. $|-5|$

31. _____

32. $|78|$

32. _____

33. $|-8.9|$

33. _____

34. $\left|-\dfrac{6}{11}\right|$

34. _____

Insert <, >, or = in the appropriate space to make a true statement.

35. $|-3|$ _____ 3

35. _____

36. $|6|$ ___ $|-7|$

36. _____

37. $|0|$ ___ $|-8|$

37. _____

38. $\dfrac{45}{9}$ ___ $\dfrac{-60}{6}$

38. _____

Tell whether the statement is true or false. If false, state why.

39. Every whole number is a real number.

39. _____

40. Every real number is a rational number.

40. _____

41. Every integer is a whole number.

41. _____

Use the following table to answer the remaining questions.

According to MSN Weather, the chart below tells us the record low temperatures for the Windy City of Chicago, IL.

Record Low Temperatures for Chicago, IL			
Month	Temperature (°F)	Month	Temperature (°F)
January	-24°	July	51°
February	-12°	August	49°
March	0°	September	37°
April	32°	October	24°
May	32°	November	0°
June	41°	December	-20°

42. Which month has the lowest temperature on record?   42. _____

43. Which month has the highest temperature on record?   43. _____

44. Put the months in order from lowest to highest temperature.   44. _____

Section 1.3 Fractions

Learning Objectives
1. Write fractions in simplest form.
2. Multiply and divide fractions.
3. Add and subtract fractions.

Vocabulary.
Use these words to complete the following sentences.

Composite	**Factors**	**Product**
Denominator	**Numerator**	**Reciprocals**
Equivalent	**Prime**	**Simplified**

1. Fractions that are on the same spot on the number line are called _____.

2. In a fraction, the top is called the _____ and the bottom is called the _____.

3. When a fraction is reduced down to it lowest terms, it is said to be _____.

4. Every _____ number can be written as a product of _____ numbers.

5. If the product of two fractions equals one, the fractions are said to be _____.

6. The result of multiplying a list of _____ is called a _____.

Objective 1.3.1
Write each number as a product of primes.

7. 16

8. 40

9. 120

10. 625

7. _____

8. _____

9. _____

10. _____

Write the fraction in lowest terms.

11. $\dfrac{35}{70}$

12. $\dfrac{18}{32}$

13. $\dfrac{24}{56}$

11. _____

12. _____

13. _____

14. $\dfrac{21}{91}$

14. _____

Objective 2

Multiply or divide as indicated. Write the answer in lowest terms.

15. $\dfrac{4}{3} \cdot \dfrac{6}{7}$

15. _____

 16. $\dfrac{2}{3} \cdot \dfrac{3}{4}$

16. _____

17. $\dfrac{8}{9} \div \dfrac{24}{27}$

17. _____

18. $2\dfrac{1}{4} \div 6\dfrac{1}{3}$

18. _____

Objective 3

Add or subtract as indicate. Write the answer in lowest terms.

19. $\dfrac{7}{8} - \dfrac{5}{8}$

19. _____

 20. $\dfrac{17}{21} - \dfrac{10}{21}$

20. _____

21. $\dfrac{15}{144} + \dfrac{27}{144}$

21. _____

Write each fraction as an equivalent fraction with the given denominator.

22. $\dfrac{5}{9}$ with a denominator of 45.

22. _____

23. $\dfrac{3}{10}$ with a denominator of $30x$.

23. _____

Add or subtract as indicated. Write the answer in lowest terms.

24. $\dfrac{4}{7} + \dfrac{5}{9}$

24. _____

25. $4 - \dfrac{10}{11}$

25. _____

26. $1\dfrac{2}{3} + 3\dfrac{2}{3}$

26. _____

Concept Extension

Solve.

27. Susie has $\dfrac{2}{3}$ teaspoons of nutmeg. Last night for dessert she used $\dfrac{1}{4}$ teaspoon. Tonight she wants to make a pumpkin pie. The recipe calls for $\dfrac{1}{2}$ teaspoon. Does Susie have enough nutmeg?

27. _____

Section 1.4   Introduction to Variable Expressions and Equations

Learning Objectives
1. Define and use exponents and the order of operations.
2. Evaluate algebraic expressions, given replacement values for variables.
3. Determine whether a number is a solution of a given equation.
4 Translate phrases into expressions and sentences into equations.

Vocabulary.
Use these words to complete the following sentences.

Add	Expression	Solution
Base	Grouping	Subtract
Exponent	Multiply	Variable

1. In the expression $2\left[3+(5-4)\right]$ you will _____ first.

2. In the expression $2\left[3+(5-4)\right]$ you will _____ second.

3. In the expression $2\left[3+(5-4)\right]$ you will _____ last.

4. A(n) _____ is a mathematical statement that two expressions are equal.

5. A numerical value that makes an equation true is the _____ .

6. Examples of _____ symbols are ( ), [ ] , and { }.

7. In the exponential expression: 4^5 , 4 is the _____ and 5 is the _____.

8. A(n) _____ is a symbol or letter that represents a number.

Objective 1
Evaluate.

9. 3^3 9. _____

10. $\left(\dfrac{4}{5}\right)^2$ 10. _____

11. 0.02^5 11. _____

Simplify each expression.

12. $3 \bullet 4 - 5 \bullet 2$ 12. _____

13. $6^2 + 12 \bullet 3 - 8 \div 4$ 13. _____

14. $2\left[5+2(8-3)\right]$ 14. _____

15. $\dfrac{18+(12-4)^2}{10^2}$ 15. _____

Evaluate each expression when x = 2, y = 3, and z = 5.

16. $4xyz$ 16. _____

17. $\dfrac{z}{3x}$ 17. _____

18. $\left| y^2 - x^3 \right|$ 18. _____

19. $\dfrac{z^3 + x^2}{y(3+z)}$ 19. _____

Objective 3
Determine whether the given number is a solution of the given equation.

20. Is 4 a solution of $\ 3x+8=5x$?   20. _____

21. Is 0 a solution of $\ x=5x+15$?   21. _____

22. Is 2 a solution of $\ 2x=4x-5$?   22. _____

Objective 4
Write each phrase as an algebraic expression. Let x represent the unknown number.

23. Two-thirds times a number increased by six.   23. _____

24. The difference of twice a number and nine.   24. _____

25. The product of three and the sum of a number and eight.   25. _____

Write each sentence as an equation or inequality. Use x to represent any unknown number.

26. The quotient of eight and a number is not equal to two.   26. _____

27. Five less a number is greater than or equal to negative seven.   27. _____

28. Twice the difference of two and a number yields nine.   28. _____

Concept Extensions

29. Explain why the square of the sum of two numbers is different than the sum of the squares of two numbers.

Section 1.5 Adding Real Numbers

Learning Objectives
1. Add real numbers with the same sign.
2. Add real numbers with unlike signs.
3. Solve problems that involve addition of real numbers.
4. Find the opposite of a number.

Vocabulary.
Use these words to complete the following sentences.

0   **-n**   **Opposites**

n   **Negative number**   **Positive number**

1. _____ are numbers that are equidistant from zero on the number line, but are on

   opposite sides.

2. The result you get from adding two opposites together is _____.

3. If n is a negative number, then $-(-n)$ is _____.

4. The result of adding -5 and 3 will result in a(n) _____ .

Objective 1
Add.

5. $64 + 739 + 165$ 5. _____

6. $-25 + (-56)$ 6. _____

7. $-4.7 + (-7.8)$ 7. _____

Objective 2
Add.

8. $86 + (-145)$ 8. _____

9. $|-6| + (-61)$ 9. _____

10. $-25 + 63 + (-45)$ 10. _____

11. $6.7 + (-3.7) + 8.9 + (-9.2)$ 11. _____

12. $[-56+41]+[78+(-102)]$ 12. _____

Objective 3
Solve.

13. The overnight low temperature in New York was $-15°F$ for Sunday night. The temperature went up $27°F$ by lunch time on Monday. What was the temperature on Monday afternoon?

13. _____

14. A frog fell down a well that was 15 feet deep. He was able to jump and land on a brick that was 3 feet from the bottom of the well. How far from the top of the well is the frog.

14. _____

Objective 4
Find each additive inverse or opposite.

15. 8   15. _____

 16. −2   16. _____

17. $|-17|$ 17. _____

Simplify each of the following.

18. $-(-15.67)$ 18. _____

19. $-|5|$ 19. _____

20. $-\left|-\dfrac{4}{7}\right|$ 20. _____

Concept Extension

21. The expression $-x + y$ can be read as "the opposite of x plus y." Write each of the following expression into words.

    a. $x + (-y)$ 21 a. _____

    b. $-x + (-y)$ 21 b. _____

    c. $-x - y$ 21 c. _____

Section 1.6 Subtracting Real Numbers

Learning Objectives
1. Subtract real numbers. 2. Add and subtract real numbers. 3. Evaluate algebraic expressions using real numbers. 4. Solve problems that involve subtraction of real numbers.

Vocabulary.
Use the choices to complete each statement.

$23 - x$	**Add**	**Opposite**
$x - 23$	**Complementary**	**Supplementary**

1. Angles whose sum is equal to 180° are called _____ angles.

2. _____ is the expression for 23 less a number.

3. Angles whose sum is equal to 90° are called _____ angles.

4. A number subtracted from 23 is the expression: _____.

5. A number decreased by 23 is the expression: _____.

6. To find the difference of two numbers you can _____ the first number to the

   _____ of the second number.

7. _____ is the expression for a number minus 23.

Objective 1
Subtract.

8. $-5 - 9$ 8. _____

9. $16 - (-3)$ 9. _____

10. $6.7 - (-12.4)$ 10. _____

11. $-\dfrac{3}{5} - \dfrac{4}{7}$ 11. _____

Perform the operation.

12. Decrease 9 by -14.

12. _____

13. Subtract 56 from 29.

13. _____

14. 15 less -31.

14. _____

Objective 2

Simplify each expression.

15. $-10-(-8)+(-4)-20$

15. _____

16. $16-24+(-32)-19$

16. _____

17. $(-2)^3 + \left[2(6-10)\right]$

17. _____

18. $|12-19|+4^2 -[8-11]^2$

18. _____

Objective 3

Evaluate each expression when $x = 2$, $y = -4$, and $z = -7$.

19. $x - z$

19. _____

20. $\dfrac{12 + y}{z - 5}$

20. _____

21. $z^2 - y^3$

21. _____

Objective 4
Solve.

22. A commercial jet liner hits an air pocket and drops 250 feet. After climbing 120 feet, it drops another 178 feet. What is its overall vertical change?

22. _____

23. Joe shot rounds of 3 over par, 2 under par, 3 under par, and 5 under par for a four day tournament of golf. What was his final score at the end of the fourth day?

23. _____

24. Sarah opened a checking account with a deposit of $101.37. She then wrote checks and had other credits against her account. What is her current balance?

Deposit		101.37
Wal-mart	65.29	
Target	54.89	
ATM	60.00	
ATM Card Fee	3.50	

24. _____

Find each unknown complementary or supplementary angle.

25.

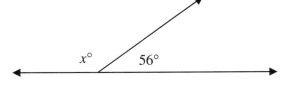

25. _____

26.

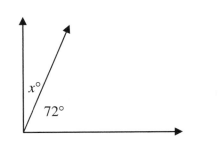

26. _____

Concept Extensions

27. If m is positive number and n is negative number, determine whether each statement is true or false. If it is false, show an example showing that it is false.

    a. $m - n$ is positive.   27 a. _____

    b. $n - m$ is positive.   27 b. _____

    c. $|m| - |n|$ is always negative.   27 c. _____

    d. $|n| - |m|$ is always a positive number.   27 d. _____

Section 1.7 Multiplying and Dividing Real Numbers

Learning Objectives
1. Multiply and divide real numbers
2. Evaluate algebraic expressions using real numbers.

Vocabulary.
Use the choices to complete each statement.

0	**Negative**	**Reciprocal**
1	**Positive**	**Undefined**

1. The result of multiplying two negative numbers will be _____.

2. When zero is the divisor, the result will be _____.

3. Another word for multiplicative inverse is _____.

4. When multiplying real numbers, if you have an odd number of negative factors, the final answer will

   be _____.

5. The quotient of a positive and a negative number will be a _____ number.

6. When zero is the dividend, the result will be _____.

7. The number _____ does not have a multiplicative inverse.

Objective 1
Multiply.

8. $-8(9)(-2)$

8. _____

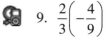 9. $\dfrac{2}{3}\left(-\dfrac{4}{9}\right)$

9. _____

10. $(-0.2)(-4.3)(-2.2)$

10. _____

11. $4(-2)(-3)(5)(-1)$

11. _____

Perform the indicated operation.

12. $(-3)(5)-4(-4)$ 12. _____

13. $8(-3)(2)+(-2)(-7)-(-12)(3)$ 13. _____

Evaluate.

14. $(-4)^3$ 14. _____

15. -2^5 15. _____

16. $(-8)^2$ 16. _____

17. -8^2 17. _____

Find the reciprocal or multiplicative inverse.

18. -14 18. _____

19. $\dfrac{5}{-6}$ 19. _____

20. $\dfrac{1}{3.4}$ 20. _____

Divide.

21. $24 \div (-4)$

21. _____

22. $\dfrac{0}{-12}$

22. _____

23. $-\dfrac{5}{9} \div \left(-\dfrac{3}{4}\right)$

23. _____

24. $64 \div 0$

24. _____

Objective 2

If $x = -4$ and $y = 3$, evaluate each expression.

25. $5x - 8y$

25. _____

26. $\dfrac{3y + 12x}{-13}$

26. _____

27. $2y^3 + y^2 - y - 6$

27. _____

Concept Extensions.

Simplify.

28. $\dfrac{\left|-8+6\right|^3+\left|-3-7\right|^2}{\left[2\left(4-3^2\right)+4\right]}$

28. _____

29. List any real numbers that are their own reciprocal.

29. _____

Section 1.8   Properties of Real Numbers

Learning Objectives
1. Use the commutative and associative properties.
2. Use the distributive property.
3. Use the identity and inverse properties.

Vocabulary.
Use the choices to complete each statement.

Additive inverses	**Associative property of multiplication**
Associative property of addition.	**Commutative property of multiplication**
Commutative property of addition	**Distributive property**
Identity element of addition	**Identity element of multiplication**

1. $6 + z = z + 6$ is an example of the _____.

2. $5(xy) = (5x)y$ is an example of the_____.

3. If the sum of two numbers is 0, they are said to be _____.

4. $9 + 0 = 9$ is an example of the _____

5. $(2 + x) + w = 2 + (x + w)$ is an example of the _____.

6. $4 \bullet h = h \bullet 4$ is an example of the _____.

7. $6 \bullet 1 = 6$ is an example of the _____.

8. $3(x - 7) = 3x + 21$ is an example of the _____.

Objective 1
Use commutative property to complete each statement.

 9. $x + 16 =$_____   9. _____

10. $6 \bullet p =$_____   10. _____

11. $5j + 3k =$_____   11. _____

Use associative property to complete each statement.

12. $5 + (d + 4) =$_____   12. _____

13. $(10c)s =$_____   13. _____

14. $(-6 + r) + t =$_____   14. _____

Use the commutative and associative properties to simplify each expression.

15. $(4+b)+9$ 15. _____

16. $8(7g)$ 16. _____

17. $-\dfrac{8}{13}\left(s\bullet\dfrac{39}{16}\right)$ 17. _____

Objective 2
Use the distributive property to write each expression without parentheses.

18. $3(x-9)$ 18. _____

19. $-(r-3-7p)$ 19. _____

20. $\dfrac{3}{4}(8x-16y)$ 20. _____

21. $5x+7(10-4x)$ 21. _____

Use the distributive property to write each sum as a product.

22. $8\bullet x+8\bullet 2$ 22. _____

23. $11x+11y$ 23. _____

24. $-5t-25$ 24. _____

25. $\dfrac{1}{4}t+\dfrac{1}{2}s$ 25. _____

Objective 3

26. Which of the following $\left\{1, -1, 0, \dfrac{5}{6}, -\dfrac{5}{6}, \dfrac{6}{5}, -\dfrac{6}{5}\right\}$ is the

    a. Reciprocal of $-\dfrac{5}{6}$ 26 a. _____

    b. Opposite of $\dfrac{6}{5}$ 26 b. _____

Concept Extensions

27. Give an example to show that division does not follow the commutative property.

           27. _____

28. Simplify the expression $16 - (12 - 9)$ and the expression $(16 - 12) - 9$ to show that subtraction does not follow the rules of the associative property.

           28. _____

Chapter 1 Vocabulary

Vocabulary Word	Definition	Example
Set	Collection of elements, enclosed in braces.	$\{1, 2, 3, ...\}$
Base	The repeated factor of an exponential expression	2^5; 2 is the base.
Denominator	The bottom number of a fraction.	$\frac{2}{5}$; 5 is the denominator.
Equation	A mathematical statement that two expressions are equal.	$3x + 4 = 16$
Exponent	The number of times that the base is repeated in an exponential expression.	2^5; 5 is the exponent
Grouping symbols		$\{\ \}, [\ ], (\ )$
Inequality symbols		$<, >, \leq, \geq, \neq$
Numerator	The top number of a fraction.	$\frac{2}{5}$; 2 is the numerator
Opposites	Two numbers that are the same distance from 0, but are on opposite sides of 0 on the number line.	18 and -18 are opposites.
Reciprocals	Two numbers whose product is 1.	$\frac{3}{4}$ and $\frac{3}{4}$ are reciprocals.
Solution	A value of a variable that makes the equation true.	$x + 7 = 19$ 12 is a solution. $12 + 7 = 19$
Variable	A symbol or letter that represents a number.	x, y, or z

Name:

Instructor:

Chapter 1 Practice Test   A

Translate the statement into symbols.

1. Eight more then twice a number is less than negative fifteen.

1. _____

 2. Subtract half a number from twenty-one.

2. _____

Simplify the expression.

3. $(-18)(-9)$

3. _____

4. $\dfrac{5}{11} - \left(-\dfrac{3}{7}\right)$

4. _____

5. $14 \div 2 \bullet (-4) + 16 \div 4$

5. _____

6. $-3\dfrac{2}{3} \bullet \left(-5\dfrac{1}{4}\right)$

6. _____

7. $\dfrac{0}{12}$

7. _____

8. $\dfrac{|12-22|}{|18-9|}$

8. _____

9. $-17 + (-13) + 19$

9. _____

Martin-Gay *Beginning and Intermediate Algebra, Fourth Edition*

10. $(7-10)^3 - 9(2)^2 + 20(-1)$

10. _____

11. $|12-19| + 3[9 - 2(5+2)]$

11. _____

Insert $<, >$, or $=$ in the appropriate space to make the statement true.

12. -18 _____ 12

12. _____

13. $|-22|$ ____ $|3(8)|$

13. _____

14. -2.756 _____ -2.75600

14. _____

15. Given the set $\left\{-12, -4, 0, \dfrac{5}{9}, 1, \pi, \sqrt{25}, 11.75\right\}$, list the numbers that also belong to the set of:

a. Natural Numbers

15 a. _____

b. Whole Numbers

15 b. _____

c. Integers

15 c. _____

d. Rational Numbers

15 d. _____

e. Irrational Numbers

15 e. _____

f. Real numbers

15 f. _____

16. Evaluate $2x^2 - 3x - 1$ if $x = -3$.

16. _____

Identify the property illustrated by each expression.

17. $0 + t = t$

17. _____

18. $(5+n) + m = 5 + (n+m)$

18. _____

19. $\dfrac{4}{9} \cdot 1 = \dfrac{4}{9}$ 19. _____

20. $-6(7-m) = -42 + 6m$ 20. _____

21. Find the multiplicative inverse of -15.

                                                               21. _____

22. Find the opposite of $-\dfrac{5}{7}$.

                                                               22. _____

Solve.

23. Over a seven day period the Dow Jones Industrial Average has gains of 32, 24, and 18 points. In that period it had loses of 29, 37, 15, and 21 points. Find the average daily performance over then seven day period.

                                                               23. _____

24. Two contractors bid on a home remodeling a house. The first bids $9,350 for the entire job. The second contractor will work for $25.50 per hour, plus $4,200 for materials. He estimates that the job will take 220 hours. Which contractor has the lower bid?

                                                               24. _____

25. A student saved $18,000 to attend graduate school. If she estimates that her expenses will be $723.50 a month while in school, does she have enough to complete her 18-month master's degree program?

                                                               25. _____

Chapter 1 Practice Test B

Translate the statement into symbols.

1. The quotient of a number and negative three is not equal to five.

   a. $\dfrac{-3}{x} \neq 5$ b. $-3x \neq 5$ 1. _____

   c. $\dfrac{x}{-3} \neq 5$ d. $x(-3) \neq 5$

2. The difference of sixteen and a number is less than twelve.

   a. $16 - x < 12$ b. $x - 16 < 12$ 2. _____

   c. $16 - x > 12$ d. $x - 16 > 12$

Simplify the expression.

3. $(-1440) \div 12$

   a. -240   b. -120   3. _____

   c. -12   d. -144

4. $\left(-\dfrac{6}{7}\right) - \left(-\dfrac{2}{9}\right)$

   a. $-\dfrac{68}{63}$ b. $-\dfrac{1}{4}$ 4. _____

   c. $-\dfrac{1}{2}$ d. $-\dfrac{40}{63}$

5. $\left[-12 \div 3(7 - 10)\right] + (-2)^3$

   a. -4   b. 4   5. _____

   c. $-\dfrac{20}{3}$ d. -20

6. $-6.724 \bullet (-3.25)$

   a. 21.853   b. -9.974   6. _____

   c. -3.474   d. 2.0689

7. $\dfrac{-5}{0}$

    a. 0   b. $\varnothing$ 7. _____

    c. -5   d. 5

8. $\left|19-4^2+(-14)\right|-\left|13-22\right|$

    a. -2   b. 2   8. _____

    c. -31   d. -9

9. $-(7)^3-(8-18)^3$

    a. -657   b. -1343 9. _____

    d. 657   d. 1343

10. $2+3\bullet\left|4-\left(7^2-6^2\right)\right|$

    a. 45   b. -25 10. _____

    c. -45 d. 29

11. $\dfrac{3^2-2\bullet4}{-30+2\bullet4^2}$

    a. $-\dfrac{1}{16}$ b. $\dfrac{1}{2}$ 11. _____

    c. 14   d. $\dfrac{1}{4}$

Insert $<$, $>$, or $=$ in the appropriate space to make the statement true.

12. -2^3 ____ $(-2)^3$

    a. $>$ b. $<$ c. $=$ 12. _____

13. $\left|6-4^2\right|$ ____ $\left|-20+2(5-2)\right|$

    a. $>$ b. $<$ c. $=$ 13. _____

14. $-\dfrac{1}{4}$____-0.245

        a. >   b. <   c. =   14. _____

15. Tell which set or sets the number -5 belongs to.

        a. Integer, Rational, Real   b. Whole, Real   15. _____

        c. Irrational, Real   d. Real

16. Evaluate $b^2 - 4ac$ if $a = -2$, $b = -5$ and $c = 7$

        a. 31   b. 81   16. _____

        c. -81 d. -31

Identify the property illustrated by each expression.

17. $-3 \bullet 1 = -3$ 17. _____

        a. Identity element of addition   b. Inverse element of addition

        c. Identity element of multiplication d. Inverse element of multiplication

18. $(5p)q = 5(pq)$ 18. _____

        a. Associative property of multiplication   b. Distributive property

        c. Commutative property of multiplication   d. Associative property of addition

19. $-(3z - 5y) = -3z + 5y$ 19. _____

        a. Associative property of multiplication   b. Commutative property of multiplication

        c. Distributive property   d. Associative property of addition

20. $1 + 0 = 1$ 20. _____

    a. Identity element of addition   b. Inverse element of addition

    c. Identity element of multiplication   d. Inverse element of multiplication

21. Find the reciprocal of $\dfrac{-13}{14}$.

    a. $\dfrac{14}{13}$ b. $\dfrac{-14}{13}$ 21. _____

    c. $\dfrac{13}{14}$ d. $-\dfrac{13}{14}$

22. Find the additive inverse of $-(-4)$

    a. 4   b. -4 22. _____

    c. $\dfrac{1}{4}$ d. $-\dfrac{1}{4}$

Solve.

23. Four students recorded time they spent working on a take-home exam: 5.2, 4.7, 9.5, and 8 hours. Find the average time spent.

    a. 6.75 hours   b. 6.85 hours

    c. 7 hours   d. 6.50 hours

                                        23 _____

24. How many pieces of pipe that are $\dfrac{2}{3}$ foot long must be laid together to make a pipe that is 16 feet long?

    a. 11 pieces   b. 24 pieces

    c. 32 pieces   d. 48 pieces

                                        24. _____

25. The expression $2L + 2W$ gives the perimeter of a rectangle with the length, L and the width, W. Find the perimeter of a rectangle whose length is $16\frac{2}{3}$ feet and width is $7\frac{3}{4}$ feet.

    a. $32\frac{5}{12}$ feet   b. $48\frac{5}{6}$ feet

    c. $24\frac{5}{12}$ feet   d. $36\frac{5}{6}$ feet

                                          25. _____

Chapter 2 Equations, Inequalities, and Problem Solving
Section 2.1 Simplifying Algebraic Expressions

Learning Objectives
1. Identify terms, like terms, and unlike terms.
2. Combine like terms.
3. Use the distributive property to remove parentheses.
4. Write word problems as algebraic expressions.

Vocabulary.
Use the choices to complete each statement.

Combine like terms   **Numerical coefficient**   **Like terms**

Terms   **Unlike terms**   **Exponent**

1. A product of a number and variables raised to powers is called a(n) _____.

2. Terms that have the same variables raised to the same power are called _____.

3. In the term $3x^2$, the three is called the _____.

4. To simplify the expression $8t + 10t$ we _____.

Objective 1

5. Complete the following table for the expression $3x^4 + \dfrac{1}{5}x^3 - x + 7$

Term	Numerical Coefficient	Variable	Exponent

6. Determine whether the following are like or unlike terms

   a. $5x$, $\dfrac{4}{25}x$ 6a. _____

   b. $17xy^2$, $27x^2y$ 6b. _____

Objective 2
Simplify each expression by combining like terms.

7. $17t + 3s - 29t - 9s$ 7. _____

8. $8x^3 + x^3 - 11x^3$

8. _____

9. $1.5x - 2.5xy + 8.8yx - 9x$

9. _____

Objective 3
Simplify each expression. First use the distributive property to remove any parentheses.

10. $6(t-5)$

10. _____

11. $-\dfrac{1}{4}(12y - 8x + 16z)$

11. _____

12. $5(x+2) - (3x-4)$

12. _____

13. $0.6(t-3.2) - (1.4 - 3.7y)$

13. _____

Objective 4

Write each of the following as an algebraic expression. Simplify if possible.

14. Add $12x - 8$ and $-16z + 19$.

14. _____

15. Subtract $3z - 9y$ from $18y - 5z$.

15. _____

Write each of the following as an algebraic expression. Simplify if possible. Let x represent the unknown number.

16. Eight times a number increased by half a number less twelve.

16. _____

17. The sum of twice a number, eighteen, and negative five times a number.

17. _____

18. The difference of a number and four, subtracted from the sum of nine and twice a number.

18. _____

Concept Extension

19. If the value of z was doubled, what would happen to the value of 17z?

19. _____

20. If both c and d were doubled, what would happen to the value of $7cd^2$?

20. _____

Section 2.2 The Addition and Multiplication Properties of Equality

Learning Objectives
1. Define linear equations and use the addition property of equality to solve linear equations.
2. Use the multiplication property of equality to solve linear equations.
3. Use both properties of equality to solve linear equations.
4. Write word phrases to algebraic expressions.

Vocabulary.

Use the choices to complete each statement.

Addition   **Equation**   **Expression**

Multiplication   **True**   **False**

1. True or false: $x = 3$ is equivalent to $3 = x$. _____

2. A(n) _____ contains an equal sign, while a(n) _____ does not.

3. $x + 2 = 9$ and $x + 2 - 2 = 9 - 2$ are equivalent equations by the _____ property of

   equality.

4. True or false. To solve for x in the equation $\dfrac{9x}{7} = 3$, the next step would look like this: $\dfrac{9x}{7} \cdot \dfrac{7}{9} = 3 \cdot \dfrac{7}{9}$.

   _____

5. By the _____ property of equality, $\dfrac{5}{9}x = 6$ and $5x = 54$ are equivalent.

Objective 1

Solve each equation.

6. $3x = 25 - 2x$ 6. _____

7. $5b - 0.7 = 6b$ 7. _____

8. $12.6 - 2.5x - 6.7 = -1.5(x - 6)$ 8. _____

Objective 2

Solve each equation.

 9. $-5x = 20$

9. _____

10. $\dfrac{-7}{9}r = \dfrac{14}{27}$

10. _____

11. $4.5x = -16.875$

11. _____

Objective 3

Solve.

12. $-x - 19 = -34$

12. _____

 13. $z - 5z = 7z - 9 - z$

13. _____

14. $-x - 3 - 8x = 2(5 - x)$

14. _____

Objective 4

Solve.

 15. Two numbers have a sum of 20. If one number is p, express the other number in terms of p.

15. _____

16. A rope is cut into 8 equal pieces. If the length of the rope was t feet initially, how long is each piece in terms of t ?

16. _____

Write each algebraic expression described. Simplify if possible.

17. If x is the first of three consecutive odd integers, express the sum of the first and third integer in terms of x.

17. _____

18. If the width of a rectangle is x and the length is the next consecutive even integer from x. What is the perimeter of the rectangle in terms of x.

18. _____

Concept Extensions

19. A football costs x and flags cost y. How much would it cost to buy 6 footballs and 16 flags to prepare for flag football season?

19. _____

20. The equation $(x+4)^2 = x^2 + 8x - 16$ has no solution. Plug in three values for x and describe what happens.

20. _____

Section 2.3 Solving Linear Equations

Learning Objectives
1. Apply a general strategy for solving a linear equation
2. Solve equations containing fractions.
3. Solve equations containing decimals.
4. Recognize identities and equations with no solutions

Objective 1

Solve each equation.

1. $-5y + 12 = 3(4y - 8)$

1. _____

2. $5(2x - 1) - 2(3x) = 1$

2. _____

3. $-(14x - 18) + 2(8x + 16) = 4(7)$

3. _____

4. $5 - (2x + 6) = -5 + 6(4x - 3)$

4. _____

5. $8(p - 2) = 5(p + 4) - p$

5. _____

Objective 2

Solve each equation.

6. $\dfrac{x}{3} + 2 = -9$

6. _____

7. $\dfrac{2x}{5}+\dfrac{3}{5}=\dfrac{9}{10}$ 7. _____

8. $\dfrac{x}{2}-1=\dfrac{x}{5}+2$ 8. _____

9. $\dfrac{5}{8}x-\dfrac{3}{4}=\dfrac{11}{12}x$ 9. _____

10. $\dfrac{5x-12}{6}=x+9$ 10. _____

Objective 3

Solve each equation.

11. $0.5x-1.2=0.35x$ 11. _____

12. $0.50x+0.15(70)=35.5$ 12. _____

13. $-0.04x+0.25(x-4)=0.5(x-2)$ 13. _____

14. $0.04(4y-2)=0.02(2+3y)+8$ 14. _____

Objective 4

Solve each equation.

 15. $2(x+3)-5=5x-3(1+x)$

15. _____

16. $3(x-3)+5x+16=4(3x-5)-2(2x+1)$

16. _____

17. $0.02x+0.03(x+4)=0.12(x+1)-0.7x$

17. _____

18. $\dfrac{1}{5}x-\dfrac{1}{2}=\dfrac{1}{20}(4x-5)-\dfrac{3}{4}$

18. _____

Concept Extension

19. Solve the equation: $0.80(x+5)=\dfrac{3}{2}(x-2)$.

19. _____

20. The perimeter of a quadrilateral is 124 inches. The sides of the quadrilateral are x, $2.5x$, $\dfrac{3}{4}(x-2)$, and $10+x$. Find the length of all four sides.

20. _____

Section 2.4 An Introduction to Problem Solving

Learning Objectives
1. Solve problems involving direct translations.
2. Solve problems involving relationships among unknown quantities.
3. Solve problems involving consecutive integers.

Objective 1

Write each of the following as equations. Then solve.

1. Twice a number increased by eight amounts to sixteen. Find the number.

    1. _____

2. Fourteen times a number increased by seven yields seven times the sum of one and twice a number. Find the number.

    2. _____

3. Twice the difference of a number and 8 is equal to three times the sum of the number and 3. Find the number.

    3. _____

4. If the sum of a number and seven is tripled, the result is eight times a number.

    4. _____

5. The sum of two and three times the difference of a number and five is equal to four times the difference a number and one.

    5. _____

Objective 2

Solve.

6. A 24 foot board is cut into two pieces so that the smaller piece measures x feet. If the longer piece is 6 feet less than twice the shorter piece, what is the length of both sides?

    6. _____

7. In 1986, the Chicago Bears beat the New England Patriots by 36 points in the Super Bowl. If a total of 56 points were scored, what was the final score of the game? (Source: National Football League)

7. _____

8. A sales clerk's sales in August were $2500 less than 2 times her sales of June. If his August sales were $9000, by what amount did her sales increase?

8. _____

9. The perimeter of a square piece of land is twice the perimeter of an equilateral triangle. (An equilateral triangle has three equal sides.) If one side of the square is 45 feet, find the length of the sides of the triangle.

9. _____

10. Two angles are supplemental if their sum is 180° The larger angle measures eight degrees more than three times the measure of the smaller angle. If x represents the measure of the smaller angle and these two angles are supplementary, find the measure of each angle.

10. _____

11. The width of a rectangular swimming pool is 13 meters less than the length and the perimeter is 98. Find it dimensions.

11. _____

12. According to People Magazine, Angelina Jolie and Halle Berry are in the top ten lists of highest paid actresses. Halle average salary is one-third the amount of Angelina. If combined they make $30,000, what is the average salary of each actress.

12. _____

13. A plumber cuts a 75 inch pipe into 3 pieces. If the longest piece is three times the smallest piece and the middle-sized piece is two more then the smallest. What is the length of all three pieces?

13. _____

Objective 3

Solve.

14. The sum of two consecutive even integers is 138. Find the integers.

14. _____

15. The measures of the angles of a triangle are 3 consecutive even integers. Find the measure of each angle.

15. _____

16. The sum of three consecutive odd integers is -96. Find the integers.

16. _____

17. The ages of three sisters are consecutive even integers. If the sum of three times the youngest, the middle sister's age, and twice the oldest is 58. Find the ages of the three sisters.

17. _____

18. Houses on Main Street have consecutive odd numbers. If the sum of two neighboring houses is 814. What are the numbers of each house?

18. _____

19. There is three consecutive even integers such that three times the second is six more than the sum of the first and third. Find the three integers.

19. _____

Concepts Extensions

20. If two lines intersect as the illustration shows, angle 1 (∠1) and ∠2, ∠3, and ∠4, are called vertical angles. Let the measure of ∠1 be any three random angle values. Compute the values of the other three angles. What did you discover?

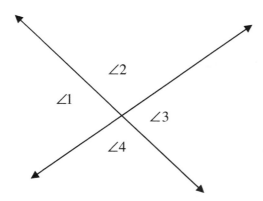

Martin-Gay *Beginning and Intermediate Algebra, Fourth Edition*

Section 2.5 Formulas and Problem Solving
Learning Objectives
   1. Use formulas to solve problems.
   2. Solve a formula or equation for one of its variables.

Objective 1

Substitute the given values into each given formula and solve for the unknown variable. If necessary, round to one decimal place.

1. $d = rt$ $d = 256$ $t = 8$ 1. _____

2. $V = l \bullet w \bullet h$ $V = 126$ $w = 3$ $h = 6$ 2. _____

3. $I = \text{P} \bullet r \bullet t$; $I = 300$, $P = 1500$, $r = 0.05$ 3. _____

4. $y = mx + b$; $y = -1$, $x = 5$, $b = -16$ 4. _____

5. $F = \dfrac{9}{5}C + 32$; $F = 77$ 5. _____

6. We are planning a trip to Disney World this year. We live 1200 miles away from Kissimmee, FL. The online map website said that it would take us about $18\dfrac{1}{2}$ hours. How fast does the website expect us to drive?

                                                   6. _____

7. Convert San Antonio Texas' highest temperature of 110 degrees Fahrenheit to Celsius. (Source: The Weather Channel)

                                                   7. _____

8. An architect designs a rectangular garden such that the width is exactly two-thirds of the length. If 260 feet of antique picket fencing are to be used to enclose the garden, find the dimensions of the garden.

8. _____

9. How long will it take an account that started with $2000 to double, if the account has a simple interest rate of 8.5%? Round your answer up to the next whole year.

9. _____

10. Joe and Janet are driving from their home in separate vehicles to the coast 425 miles away. Jeff leaves at 11:00 AM and averages 55 miles per hour. Janet leaves two hours later at 1:00 PM, but she averages 65 mph. Which person arrives at the coast first?

10. _____

11. If the perimeter of a tennis court is 228 feet and the width is three less than half the length. What are the dimensions of the tennis court?

11. _____

Objective 2

Solve each formula for the specified variable.

12. $A = l \bullet w$ for w.

12. _____

13. $A = \dfrac{1}{2}bh$ for b.

13. _____

14. $C = 2\pi r$ for r

14. _____

 15. $S = 2\pi rh + 2\pi r^2$ for h.

15. _____

16. $F = \dfrac{9}{5}C + 32$ for C.

16. _____

17. $V = \dfrac{1}{3}\pi r^2 h$ for h.

17. _____

18. $A = \dfrac{1}{2}(B + b)$ for b.

18. _____

Concept Extensions

19. Suppose that your fellow classmate analyzes a distance problem that follows "Sarah has traveled 45 miles before Mary starts. Because Mary travels 5 mph faster than Sarah, it will take her $\dfrac{45}{5} = 9$ hours to catch Sarah." How would you react to this analysis of the problem?

19. _____

20. When a car of mass m collides with a wall, the energy of the collision is given by the formula $E = \dfrac{1}{2}mv^2$. Compare the energy of two collisions: a car striking a wall at 20 mph, and at 40mph.

20. _____

Section 2.6 Percent and Mixture Problem Solving

Learning Objectives
1. Solve percent equations.
2. Solve discount and mark-up problems.
3. Solve percent increase and percent decrease problems.
4. Solve mixture problems.

Objective 1

Find each number described.

1. What number is 120% of 50?

1. _____

2. The number 45 is 25% of what number?

2. _____

3. 160% of what number is 144?

3. _____

4. The number 56.25 is 45% of what number?

4. _____

5. The number 28.6 is what percent of 52?

5. _____

6. 23% of 20 is what number?

6. _____

Objective 2

Solve. If needed round answers to the nearest cent.

7. The dress you have been looking for is on sale for 35% off of the original price. If it originally is $120.00, what is the sale price?

7. _____

8. The college bookstore will always mark-up the price of textbooks 15% of what the amount they purchased the books from the publishers. If the publisher sells a specific text to the bookstore for $85, what will the bookstore sell that text to the students for?

8. _____

9. Find the rate of discount if the discount is $50 and the original price was $400.

9. _____

10. A 18% discount on a big screen TV saved Tony Williams $154.26. What was the original price of the TV?

10. _____

11. A birthday celebration meal is $40.50 including tax. Find the total cost if 15% tip is added to the cost.

11. _____

12. Greg purchased a Jeep Cherokee. The selling price plus the 8.25% state sales tax was $34,640.00. How much tax was added to the price of the vehicle?

12. _____

Objective 3

Solve. If necessary round to the nearest hundredth.

13. The value of my house has gone up to 110,000 from 99,000. What was the percent increase?

13. _____

14. The price of gas has gone up 53% in the past three years. If three years ago the price of gas was $1.89 per gallon, what is the current price of gas?

14. _____

15. Every year we are guaranteed to get a 5% pay raise. If my contract for next year states I make $56,000, what is my salary for this year?

15. _____

16. It has been stated that the crime rate in your neighborhood is 5% lower this year than it was last year. According to records, last year there were 120 crimes committed in your neighborhood. How many crimes were committed this year?

16. _____

Objective 4

Solve. If necessary, round your answer to the nearest tenth.

17. Walnuts sell for $4.80 per pound, and pecans sell for $6.40 a pound. How many pounds of pecans should be mixed with 20 pounds of walnuts to get a mixture that can be sold for $5.40?

17. _____

18. Hawaiian Punch is 10% fruit juice. How much water would you have to add to one gallon of Hawaiian Punch to get a drink that is 6% fruit juice for your infant son.

18. _____

Martin-Gay *Beginning and Intermediate Algebra, Fourth Edition*

19. How many liters of a 20% alcohol solution should Jerry mix with 50 liters of a 60% alcohol solution to obtain a 30% solution?

19. _____

20. Gus has on hand a 5% alcohol solution and a 20% alcohol solution. He needs 30 liters of a 10% alcohol solution. How many liters of each solution should he mix together to obtain the mixture he desires?

20. _____

Concept Extensions

21. To calculate a 15 % tip on a $24 bill a customer adds $2.40 and half of $2.40, or $1.20, to get $3.60. Is this method valid? Why or why not?

21. _____

22. When a discount of 15 % is given by a retailer is it the same as giving three successive discounts of 5% each? Give arguments to support your answers.

22. _____

Section 2.7 Further Problem Solving

Learning Objectives
1. Solve problems involving distance.
2. Solve problems involving money.
3. Solve problems involving interest.

Objective 2.8.1

Solve. Round answers to once decimal point if necessary.

1. Two cars left St. Louis at the same time. One traveled east at 60 mph and the other went west at 50 mph. How long until the two cars are 715 miles apart?

1. _____

2. Ashley and Brandon are 315 miles apart. Ashley hops in her car and drives 50 mph towards Brandon. At the same time, Brandon hopped in his car and drove toward Ashley. He drove at a rate of 55 mph. How long does it take them to meet?

2. _____

3. Suppose that Carl was riding a bicycle at 15 mph, he rode 10 miles farther than Michael, who was riding his bike at 14 mph. Carl rode for 30 minutes longer than Michael. How long did Carl ride his bike?

3. _____

4. How long will it take a bus traveling at 60 miles per hour to overtake the car traveling at 40 mph if the car had a 1.5 hour head start?

4. _____

5. Two cars start together and head east, one averaging 56 mph and the other averaging 67 mph. How far apart will the cars be in $7\frac{1}{2}$ hours?

5. _____

Martin-Gay Beginning and Intermediate Algebra, Fourth Edition

6. On a Sunday morning, Erika walked for 2 hours and then ran for 30 minutes. If she ran twice as fast as she walked and she covered 12 miles altogether, then how fast did she run?

6. _____

Objective 2

Solve. Round to the nearest tenth if needed.

7. Bobby has 75 coins in his piggy bank. He had only nickels and dimes. If he counted his money and the total was $4.50, how many of each coin does he have?

7. _____

8. Part of the proceeds from a garage sale was $280 worth of $5 and $10 bills. If there were 20 more $5 bills than $10 bills, find the number of each denomination.

8. _____

9. The manager of an electronics store placed an order for $21,320 worth of DVD players at $160 each and MP3 players at $120 each. If the number of MP3 players that were ordered was three times the number of DVD players, then how many of each did the manager order?

9. _____

10. The cost for a long-distance phone call is $0.45 for the first minute and $0.35 for each additional minute. If the total charge for a long-distance call is $6.40, how many minutes was the call?

10. _____

11. Tabitha is taking guitar lessons. Because she was not dedicated in the beginning of her lessons, Tabitha's parents make her earn the money to pay for her lessons. Her parents pay her $1.25 to do the laundry and $5.00 to mow the yard. In one month, she did the laundry 8 times as much than she mowed the yard. If her parents paid her $45.00 that month, how many times did she mow the yard?

11. _____

12. Danny works part time as a cashier making $7.25 an hour. During one week he worked 32 hours and made $243.84. He realizes he must have gotten a raise that he did not know about. How much per hour was his raise?

12. _____

Objective 3

Solve. If necessary round to the nearest tenth.

13. Frank invests a certain amount of money at 5% interest and $1360 more than that amount at 7%. His total yearly interest was $241.60. How much did she invest at each rate?

13. _____

14. How can $54,000 be invested, part at 8% annual simple interest and the remainder at 10% annual simple interest, so that the interest earned by the two accounts will be equal?

14. _____

15. A sum of $8000 is invested, part of it at 8% and the remainder at 10%. If the interest earned by the 8% investment is $260 less than the interest earned by the 10% investment, find the amount invested at each rate.

15. _____

16. If $4000 is invested at 5% interest, how much should be invested at 8% interest so that the total return for both investments averages 7%?

16. _____

17. How can $8400 be invested, part at 9% and the remaining at 12%, so that the two accounts will produce the same amount of interest?

17. _____

18. Wayne makes $8000 more per year than his wife. Wayne saves 10% of his income for retirement, and his wife saves 8%. If together they save $6416 per year, then how much does each make in one year?

18. _____

Concept Extensions.

Solve.

19. How many liters of pure alcohol must be added to 30 liters of 10% solution to obtain a 30% solution?

19. _____

20. Michelle has $3.90 worth of nickels, dimes and quarters. The number of dimes is 3 less than the number of nickels. The number of quarters is 7 more than the number of dimes. How many of each coin does she have?

20. _____

Section 2.8 Solving Linear Inequalities

Learning Objectives
1. Define linear inequality in one variable, graph solution sets on a number line, and use interval notation.
2. Solve linear inequalities.
3. Solve compound inequalities.
4. Solve inequality applications.

Vocabulary.
Use the choices to complete each statement.

Linear inequality in one variable   **True**
Compound inequality   **False**

1. True or false. When dividing both sides of an inequality by a negative number, the inequality remains

   the same. _____

2. A _____ contains two inequalities.

3. $ax + b < c$ is the general form of a _____.

Objective 1

Graph each set of numbers given in interval notation. Then write the inequality statement in x describing the numbers graphed.

4. $(3, \infty)$ 4. _____

5. $(-\infty, -1]$ 5. _____

6. $[10, \infty)$ 6. _____

Graph each inequality on a number line. Then write the solutions in interval notation.

 7. $x \leq -1$ 7. _____

8. $y > -3$ 8. _____

9. $w \geq 6$ 9. _____

Objective 2

Solve each inequality. Graph the solution set and write it in interval notation.

10. $4x < 16$ 10. _____

11. $x - 3 \geq -8$ 11. _____

12. $-8x \leq 16$ 12. _____

13. $4x - 6 < 3x + 1$ 13. _____

14. $8(5 - x) \leq 10(8 - x)$ 14. _____

15. $\dfrac{3x - 3}{2} < 2x + 2$ 15. _____

Objective 3

Solve each inequality. Graph the solution set and write it in interval notation.

16. $-4 < x \le 7$ 16. _____

17. $-12 \le x - 6 < 9$ 17. _____

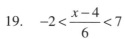 18. $-6 < 3(x-2) \le 8$ 18. _____

19. $-2 < \dfrac{x-4}{6} < 7$ 19. _____

Objective 4

Solve.

20. A student has test scores of 70, 89, 76, and 82 points. What must the student get on her fifth test so that her average is at least 80 points?

20. _____

21. Two less three times a number is less than six more than the number.

21. _____

22. Find the values of x so that the perimeter of this rectangle is no greater than 100 centimeters.

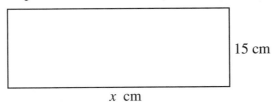

15 cm

x cm

22. _____

23. A bride does not want to spend more than $3000 on her wedding reception. If it cost $20 per plate, how many people can she invite to the wedding reception and not go over budget?

23. _____

Concept Extensions.

24. Given the inequality $1 < \dfrac{1}{x}$; is it always true that $x < 1$? Why or why not?

24. _____

Chapter 2 Vocabulary

Vocabulary Word	Definition	Example
Like terms	Terms with the same variable raised to the same power.	$5x$, $9x$, and $-4x$ are like terms.
Numerical Coefficient	The numerical factor of a term	$2x$ 2 is the numerical coefficient
Linear inequality in one variable	An inequality in the form $ax + b < c$ (The $<$ can be replaced by $>$, $\leq$ or $\geq$.	$3x - 6 < 12$
Equivalent equations	Equations that have the same solution.	$x + 2 = 4$ $2x + 6 = 10$
Formula	An equation that describes a known relationship among quantities	$d = rt$
Compound inequality	Inequalities that have two inequality symbols	$-3 < x < 9$
Linear equation in one variable	An equation in the form $ax + b = c$, where a, b, and c are all real numbers.	$3x + 9 = 18$

Martin-Gay *Beginning and Intermediate Algebra, Fourth Edition*

Chapter 2 Practice Test A

Simplify each of the following expressions.

1. $8t - 19 + (-12t) + 37$

1. _____

2. $3(x - 7) - (2x + 9)$

2. _____

3. $1.4(3.2p - 8) + 4.6 - 7.8$

3. _____

4. $\dfrac{1}{2}x + \dfrac{3}{7}\left(\dfrac{1}{4}x - 6\right)$

4. _____

Solve each of the following equations.

5. $5x + 13 = -3^3$

5. _____

6. $3\left[4x - (5x + 12)\right] = 24$

6. _____

7. $\dfrac{2}{3}x - \dfrac{4}{5} = \dfrac{1}{2}x$

7. _____

8. $0.06x - 0.14 = 0.2(x + 7)$

8. _____

9. $7y - 34 + 9y = 4^2(y - 2)$

9. _____

10. $4(x - 6) - 3(x + 5) = 2(x + 3) + 2(x - 6)$

10. _____

11. $\dfrac{2x - 3}{9} + \dfrac{x + 1}{2} = x - 4$

11. _____

12. $0.08(t - 200) = 44.2 - 0.06t$

12. _____

13. $23 + 4(n - 2) = 9n + 5(3 - n)$

13. _____

Solve each of the following applications.

14. Find the value of B if $b = 19$, and $A = 26$ in the formula . $A = \dfrac{1}{2}(B+b)$

14. _____

15. The quotient of three times a number and 5 amounts to half the number increased by two.

15. _____

16. Find three consecutive odd integers such that the sum of the smallest and four times the largest is 61.

16. _____

17. How much 35% copper alloy should be melted with 37 kg of 75% copper alloy to produce an alloy which is 50% copper. If necessary round your answer to the nearest tenth.

17. _____

18. A bus leaves the station and travels at 50 mph. An hour and a half later, a second bus leaves that same station, but travels at 65 mph. When does the second bus overtake the first?

18. _____

19. The perimeter of a standard football field, including the end zones is 1040 feet. If the field is 360 feet long, how wide is the field?

19. _____

Solve each of the following equations for the indicated variable.

20. $S = P + \mathrm{Pr}t$ for t.

20. _____

21. $y = mx + b$ for m.

21. _____

Solve and graph each of the following. Write your answers in interval notation.

22. $2x + 8 > 17$

22. _____

23. $3 - \dfrac{1}{4}x \le 2$

23. _____

24. $2 \le 4 - \dfrac{1}{2}(x - 8) \le 10$

24. _____

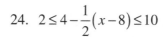

Solve.

25. A rectangle's width is three feet more the half its length. If the perimeter of the rectangle is between 12 and 21 feet, what must the measurement of the length be between?

25. _____

Chapter 2 Practice Test B

Simplify each of the following expressions.

1. $7.3r - 2.3 - 3.4r + 6.7 + 2r$ 1. _____

    a. $5.9r - 4.4$ b. $1.9r + 11$

    c. $5.9r + 4.4$ d. $12.9r - 11$

2. $4(t - 3) - 5(7 - t)$ 2. _____

    a. $9t - 23$ b. $-t + 23$

    c. $-t - 47$ d. $9t - 47$

3. $-\left[17 + 3(2x - 8) - 4(-x)\right]$ 3. _____

    a. $-10x + 7$ b. $-2x + 7$

    c. $10x - 7$ d. $-2x - 41$

4. $\dfrac{5}{6}(x - 2) - \dfrac{7}{8}(x + 4)$ 4. _____

    a. $1\dfrac{17}{24}x + 1\dfrac{5}{6}$ b. $-\dfrac{1}{24}x - 5\dfrac{1}{6}$

    c. $-\dfrac{1}{24}x + 1\dfrac{5}{6}$ d. $-\dfrac{1}{24}x + 5\dfrac{1}{6}$

Solve each of the following equations.

5. $\dfrac{1}{7} + x = -4$ 5. _____

    a. $-\dfrac{29}{7}$ b. $\dfrac{27}{7}$

    c. $\dfrac{29}{7}$ d. $-\dfrac{27}{7}$

6. $6g + 5 - 2(g + 2) = -(-7g - 1)$ 6. _____

    a. $\dfrac{2}{3}$ b. 0

    c. -10 d. $-\dfrac{8}{15}$

7. $\dfrac{1}{3}(15x - 6) = 20\left(\dfrac{2}{5}x - \dfrac{1}{4}\right) - 3x$ 7. _____

    a. $\dfrac{7}{16}$ b. $\dfrac{1}{2}$

    c. all real numbers   d. no solution

8. $3(1 - 6h) = -2(3h + 1) + 9$ 8. _____

    a. $-\dfrac{1}{3}$ b. $\dfrac{7}{6}$

    c. $\dfrac{1}{6}$ d. $-\dfrac{2}{3}$

9. $3a + 2\left[2 + 3(a - 1)\right] = 2(4a - 1) + a$ 9. _____

    a. 1 b. 2

    c. all real numbers   d. no solution

10. $\dfrac{4x + 5}{3} = \dfrac{2x + 6}{5}$ 10. _____

    a. $\dfrac{15}{2}$ b. $-\dfrac{1}{2}$

    c. $\dfrac{43}{14}$ d. $-\dfrac{43}{14}$

11. $0.15(x-2)-0.4(x+1)=0.1(25)$ 11. _____

    a. -1.8 b. -3.8

    c. -1.4 d. -12.8

12. $-2\left[3-(7+2x)\right]-(5-x)=7x$ 12. _____

    a. -1.3 b. -2.5

    c. 1.5 d. 0.30

13. $0.5x-1.2=0.35$ 13. _____

    a. -8 b. $\dfrac{24}{17}$

    c. 8 d. $-\dfrac{24}{17}$

Solve each of the following applications.

14. Find the value of r if $C=113.04$, for the equation $C=2\pi r$. Use 3.14 for π.

    a. 20   b. 36

                                              14. _____

    c. 18   d. 24

15. Three times the sum of a number and 2, decreased by five times a number is zero.

    a. 1   b. 3

                                              15. _____

    c. 0.50   d. -3

16. A $54 shirt is on sale for 25% off. If the sales tax is 8%, what would the total of the purchase?

    a. $15.75   b. $13.50

                                                16. _____

    c. $43.74   d. $40.50

17. I have invested a sum of money into an account that gives 5.5% interest. I invested $250 more in
     my account that gives 7% interest. If the total interest at the end of the year is $800, how much did
     I invest altogether?
     a. $6260   b. $6290

                                                                           17. _____
     c. $12,770   d. $12,830

18. There was an accident at the candy factory, and two types of candies were mixed together. There
     were 20 pounds of the chocolate chips that sell for $1.50 per pound and 40 pounds of the peanut
     butter chips that sell for $2.50 per pound. How much should the candy store sell the mixture so that
     the owners will not lose money?
     a. $5.00   b. $2.17

                                                                           18. _____
     c. $ 4.23   d. $ 3.72

19. Aaron has $6.25 in quarters and dimes. He has 10 more dimes than quarters. How many quarters
     does he have?
     a. 15   b. 5

                                                                           19. _____
     c. 13   d. 30

Solve each of the following equations for the indicated variable.

20. $A = \dfrac{1}{2}bh$ for b.

     a. $\dfrac{2A}{h} = b$ b. $\dfrac{A}{2h} = b$

                                                                           20. _____

     c. $\dfrac{Ah}{2} = b$ d. $\dfrac{2h}{A} = b$

21. $P = 2L + 2W$ for L.

     a. $P - W = L$ b. $\dfrac{P-W}{2} = L$

                                                                           21. _____

     c. $P - 2W = L$ d. $\dfrac{P-2W}{2} = L$

Solve each of the following. Write your answers in interval notation.

22. $2x > 6x - 24$

22. _____

    a. $(6, \infty)$ b. $[6, \infty]$

    c. $(-\infty, 6)$ d. $(-\infty, 6]$

23. $4(p - 4) \le 7p - 8$

23. _____

    a. $\left(-\dfrac{8}{3}, \infty\right)$ b. $\left[-\dfrac{8}{3}, \infty\right)$

    c. $\left(-\infty, -\dfrac{8}{3}\right)$ d. $\left(-\infty, -\dfrac{8}{3}\right]$

24. $-12 \le 6(x - 8) < 30$

24. _____

    a. $(6, 13)$ b. $[6, 13)$

    c. $(-6, 13)$ d. $(-6, 13]$

25.   The three sides of a triangle are given to be three consecutive integers. If the perimeter is greater than 60 but less than 120, what must the requirements for the smallest side be?

    a. $(18, 38)$ b. $(17, 37)$

25. _____

    c. $(20, 40)$ d. $(19, 39)$

Chapter 3 Graphs and Introduction to Functions
Section 3.1 Reading Graphs and the Rectangular Coordinate System

Learning Objectives
1. Read bar and line graphs.
2. Define the rectangular coordinate system and plot ordered pairs of numbers.
3. Graph paired data to create a scatter diagram.
4. Determine whether an ordered pair is a solution of an equation in two variables.
5. Find the missing coordinate of an ordered pair solution, given one coordinate of the pair.

Vocabulary.
Use the choices to complete each statement.

False	**Four**	**Origin**
Quadrants	**Solutions**	**True**
x-axis	**x-coordinate**	**y-axis**
y-coordinate	**(0, 0)**	**(10, 10)**

1. The rectangular coordinate system is created with a vertical line called the _____ and a horizontal line called the _____.

2. The rectangular coordinate system is separated into _____ regions that we refer to as _____.

3. A special point found at the intersection of the two axes is called the _____. Its ordered pair is _____.

4. True or false. Any ordered pair represents one point in the plane. _____.

5. In an ordered pair, the first number is the _____ and the second value in the pair is the _____.

6. The _____ of an equation in two variables is the ordered pair that makes the equation a true statement.

Objective 1

The following bar graph contains the percent of public school instructional rooms with internet access. (Source: National Center for Education Statistics) Use this bar graph to answer the following questions.

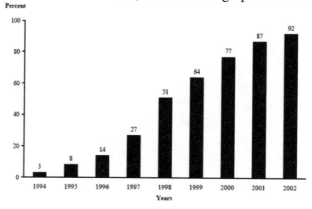

7. In between what years was there the greatest increase?

7. _____

8. What was that difference?

8. _____

9. By looking at this bar graph, can you see a relationship between the two given variables? What do you see?

9. _____

The following line graph contains the number of "unruly passengers" according to FAA regulations. Use this line graph to complete the following questions.

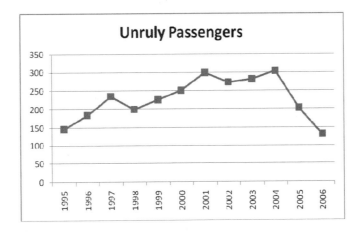

10. In what year were there the least unruly passengers?

10. _____

11. About how many unruly passengers was there in 2000?

11. _____

12. Is there a trend between the years and the number of unruly passengers? Does the year affect the number of unruly passengers?

12. _____

Objective 2

Plot each ordered pair. State in which quadrant or axis each point lies.

13. A. $(5,7)$
     B. $(-4,9)$
     C. $(9,-5)$
     D. $(-3.5,0)$
     E. $\left(0,\dfrac{12}{5}\right)$

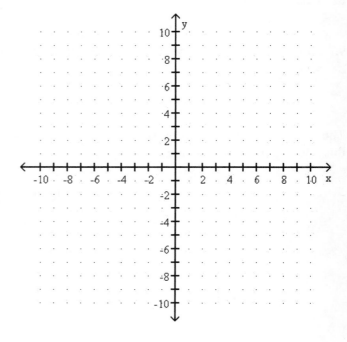

Objective 3

14. The table shows the percentage of 11th graders that said they used their computers to learn something. (Source: National Center of Education Statistics.)

Year	Percentage
1984	54.6
1988	65.3
1990	64.5
1992	72.3
1994	70.7
1996	80.2

a. Write each paired data as an ordered pair of the form (year, percentage)

14 a. _____

b. Draw a grid and create a scatter diagram of the paired data.

Objective 4

Determine whether each ordered pair is a solution of the given linear equation.

15. $2x + y = 7$ $(3,1)$ $(7,0)$ $(0,7)$

15. _____

16. $-3y + 4x = -12$ $(3,0)$ $(2,-1)$ $\left(-2, \dfrac{4}{3}\right)$

16. _____

Objective 5

Complete each ordered pair so that it is a solution of the given linear equation.

17. $x - 4y = 4$ $(\ \ ,-2)$ $(4,\ \ )$

17. _____

18. $2y = \dfrac{1}{3}x - 4$ $(-9,\ \ )$ $(\ \ ,3)$

18. _____

Complete the table of ordered pairs for each linear equation.

19. $y = -4x$

x	y
0	
	-8
- 3	

20. $y = \dfrac{1}{2}x - 5$

x	y
0	
	0
- 6	

Concept Extensions

21. Three vertices of a rectangle are $(3.2, 5.4)$, $(3.2, -2.2)$, and $(-1.6, 5.4)$

    a. Find the coordinate of the fourth vertex.

                                             21 a. _____

    b. Find the perimeter of the rectangle.

                                             21 b. _____

    c. Find the area of the rectangle.

                                             21 c. _____

Section 3.2 Graphing Linear Equations

Learning Objectives
1. Identify linear equations.
2. Graph a linear equation by finding and plotting ordered pair solutions.

Objective 1

Determine whether the given equation is a linear equation in two variables.

1. $1.4x - 2.8y = 6.4$ 1. _____

2. $y = -3$ 2. _____

3. $\frac{2}{3}x - \frac{1}{7}y = -2$ 3. _____

4. $x + y^4 = -1$ 4. _____

Objective 2

For each equation, find three ordered pair solutions by completing the table. Then use the ordered pairs to graph the equation.

5. $x + y = 5$

x	y
0	
	0
2	

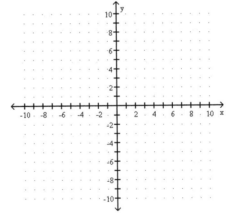

6. $y = \frac{1}{3}x - 4$

x	y
0	
6	
-9	

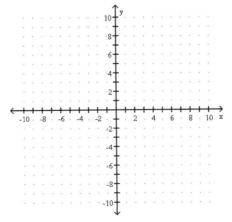

7. $y = -3x - 2$

x	y
0	
1	
2	

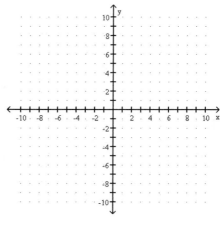

Graph each linear equation.

8. $x + y = -3$

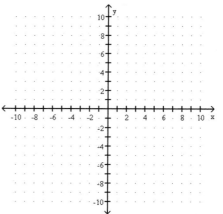

 9. $x - 2y = 6$

Martin-Gay *Beginning and Intermediate Algebra, Fourth Edition*

10. $y = -2$

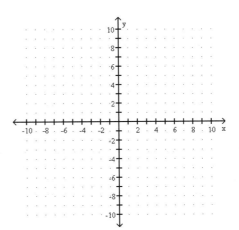

11. $2x - 4y = -8$

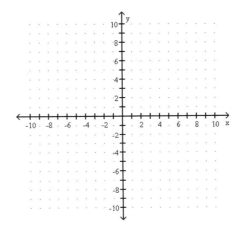

12. $y = 0.25x - 4$

13. $y = \dfrac{1}{2}x + 2$

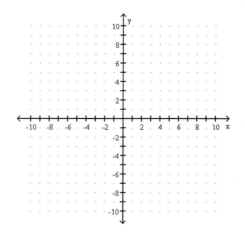

14. Graph each pair of linear equations on the same set of axes. Discuss how the graphs are similar and how they are different.

$y = 3x$ $\qquad$ $y = 3x - 4$

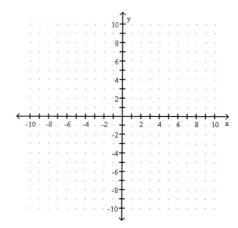

15. $y = -2x + 3$ $\qquad$ $y = 2x + 3$

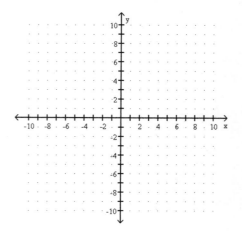

Solve.

16. The cost for a company to produce videos is $1.75 per video plus $1200 to run the equipment each month. The equation for costs of each month is $C = 1.75V + 1200$.   How much would it cost the company to make 1000 videos?

16. _____

17. If the company from problem 16, decides to charge $19 for each video, the equation for profit is $P = 19V - 1200$. How much profit would the company make if they made 1000 videos?

17. _____

Concept Extensions

18. Take the company from problem 16 and 17. Given the two equations. $P = 19V - 1200$ and $C = 1.75V + 1200$.   A break-even point for a company is the number of units that needs to be sold in order for the company to neither make a profit nor lose money. A break-even point is where the profit will equal the cost. For this company is there a break even point? Round your answer up to the next whole number of videos.

18. _____

Section 3.3 Intercepts

Learning Objectives
1. Identify intercepts of a graph.
2. Graph a linear equation by finding and plotting the intercepts.
3. Identify and graph vertical and horizontal lines.

Vocabulary.
Use the choices to complete each statement.

x	y	Horizontal
Linear	Standard	Vertical
x-intercept	y-intercept	

1. The graph $y = 2$ is a graph of a _____ line.

2. A point on the graph where x = 0 is called the _____.

3. The _____ form is $Ax + By = C$.

4. The point on the graph where it crosses the x-axis is called the _____.

5. The graph $x = -0.5$ is a graph of a _____ line.

6. Any equation that can be written as $Ax + By = C$ is called a _____ equation in two

    variables.

Objective 1

Identify the x- and the y-intercepts.

1.

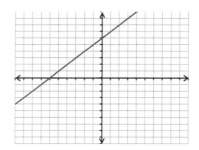

1.

2.

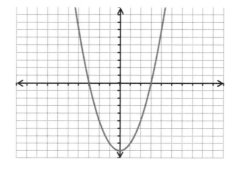

2.

Martin-Gay *Beginning and Intermediate Algebra, Fourth Edition*

Objective 2

Graph the equations by finding and plotting intercepts.

3. $2x - 3y = 12$

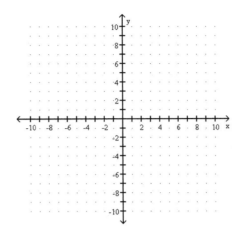

4. $-x + 2y = 6$

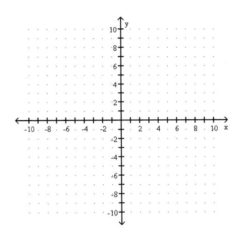

5. $y = \dfrac{1}{2}x + 3$

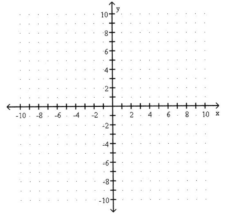

6. $y = -0.25x + 2$

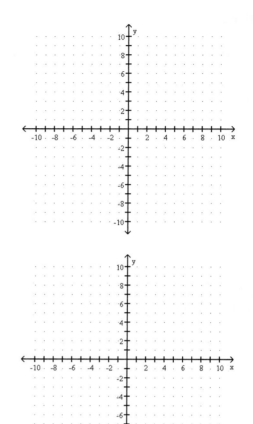

7. $y = -2x$

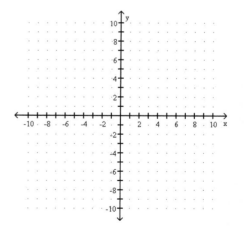

8. $4x + 3y = 5$

Martin-Gay *Beginning and Intermediate Algebra, Fourth Edition*

Objective 3

Graph each linear equation.

9. $y = 5$

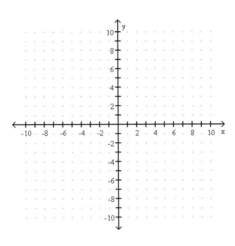

10. $y + 3 = 7$

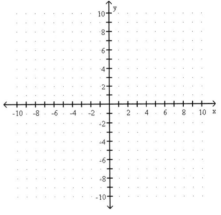

11. $x = 0$

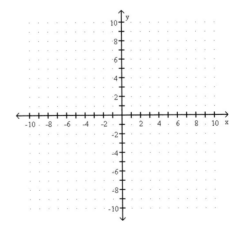

12. $x + 4 = 0$

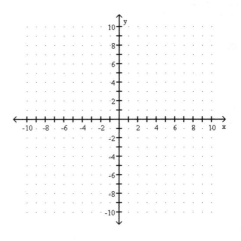

Concept Extensions

13. Do all graphs have x- and y-intercepts?

14. Are $x + y = -4$ and $-x - y = 4$ the same line? How do you know this?

Section 3.4   Slope and Rate of Change

Learning Objectives
1. Find the slope of a line given two points of the line.
2. Find the slope of a line given its equation.
3. Find the slopes of horizontal and vertical lines.
4. Compare the slopes of parallel and perpendicular lines.
5. Slope as a rate of change.

Vocabulary.

Use the choices to complete each statement.

b	*m*	**0**
Negative	**Parallel**	**Perpendicular**
Positive	**Slope**	**Undefined**

1. In the equation $y = mx + b$, _____ is the slope and _____ is the y-intercept.

2. If the value of y increases as the value of x increases, the line has a _____ slope.

3. A vertical line has a slope of _____.

4. The _____ of a line is the ratio of vertical change to its horizontal change.

5. Two lines are said to be _____ if they have the same slope, but different intercepts.

6. A horizontal line has a slope of _____.

Objective 1

Find the slope of the line that passes through the given points.

7. $(-1, 5)$ and $(6, -2)$

7. _____

8. $(2, -3)$ and $(4, -1)$

8. _____

9. $(0, -7)$ and $(0, 3)$

9. _____

10. $(-4, 9)$ and $(-11, 18)$

10. _____

Objective 2

Find the slope of the given lines.

11. $y = \frac{2}{3}x - 7$

11. _____

 12. $2x + y = 7$

12. _____

13. $4x - 5y = -9$

13. _____

Objective 3

Find the slope of the given line.

14. $x = -7$

14. _____

15. $y = 24$

15. _____

Objective 4

Determine whether each pair of lines is parallel, perpendicular or neither.

 16.

$$y = \frac{2}{9}x + 3$$

$$y = -\frac{2}{9}x$$

16. _____

17.

$3x + 4y = 7$

$6x - 8y = -3$

17. _____

18.

18. _____

$$y = -\frac{4}{9}x - 2$$
$$9x - 4y = -3$$

Objective 5

19. The pitch of a roof rises 6.5 feet over a horizontal distance of 13 feet. What is the slope of the roof?

19. _____

20. As a car drives 20 miles, the elevation of the road decreases by 5 feet. What is the grade of the road?

20. _____

Concept Extensions

21. The percentage of women who had an infant under the age of 1 and were in the labor force in 1976 was 31%. In 1998, the percentage was 59%.

   a. Rewrite this data as two ordered pairs, where x is the number of years since 1976.

21 a. _____

   b. Find the equation of the line in slope intercept form.

21 b. _____

   c. If the trend stays the same, what would the percentage be in 2007.

21 c. _____

22. What year would the percentage be 100%? Round your answer to the next whole year.

22. _____

Section 3.5   Equations of Lines

Learning Objectives
1. Use the slope-intercept form to write an equation of the line.
2. Use the slope-intercept form to graph a linear equation.
3. Use the point-slope form to find an equation of a line given its slope and a point of the line.
4. Use the point-slope form to find an equation of a line given two points of the line.
5. Find the equations of vertical and horizontal lines.
6. Use the point-slope form to solve problems.

Vocabulary.
Use the choices to complete each statement.

Horizontal	**Point-slope**	**Slope-intercept**
Standard	**Vertical**	

1. The form $y - y_1 = m(x - x_1)$ is known as _____ form.

2. If the equation of a line is in the form y = constant, it is a _____ line.

3. The form $y = mx + b$ is referred to as _____ form.

4. An example of _____ form is $3x + 4y = 9$.

5. An equation of the form x = constant, is a _____ line.

Objective 1

Write the equation of the line with each given slope, *m*, and y-intercept $(0, b)$. Write the answer in slope-intercept form.

6. $m = 2, \quad b = 4$

6. _____

7. $m = \dfrac{2}{3}, \quad b = -2$

7. _____

8. $m = 0, \quad b = -1$

8. _____

Objective 2

Use the slope-intercept form to graph each equation.

9. $y = 2x - 5$

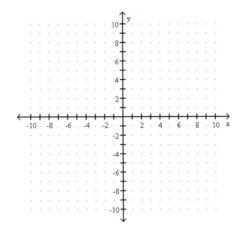

10. $y = -\dfrac{1}{4}x + 5$

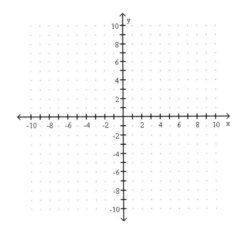

11. $4x - 7y = -14$

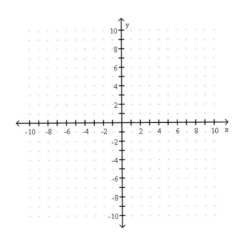

Objective 3

Find the equation of each line with the given slope that passes through the given point. Write the equation in the standard form.

12. $m = 3 \ (4, 3)$

12. _____

 13. $m = -8 \ (-1, -5)$

13. _____

14. $m = -\dfrac{1}{4} \ (-2, -4)$

14. _____

Objective 4

Find an equation of the line passing through each pair of points. Write your answer in standard form.

15. $(1, -2)$ and $(-3, 4)$

15. _____

16. $(0, -4)$ and $(7, 2)$

16. _____

17. $(-4, 2)$ and $(0, -5)$

17. _____

Objective 5

Find the equation of each line.

18. Horizontal line through $(-12,18)$.

18. _____

19. Vertical line through $(24,-16)$

19. _____

20. Parallel to the x-axis, through $(2,-3)$

20. _____

21. Perpendicular to $y = 2$, through $(-7,9)$

21. _____

Objective 6

Solve. Assume the exercise describes a linear relationship. Write the equation in slope intercept form.

22. A rock is dropped from the top of a 400-foot cliff. After 1 second, the rock is traveling 32 feet per second. After 3 seconds, the rock is traveling 96 feet per second.

22. _____

Concept Extensions

23. Suppose there is a line with a slope of $\dfrac{2}{3}$ and contains the point $(4,6)$. Are the points $(10,7)$ and $(-8,-2)$ also on the line? What is the equation of the line in slope-intercept form?

23. _____

Section 3.6 Functions

Learning Objectives
1. Identify relations, domains, and ranges.
2. Identify functions.
3. Use the vertical line test.
4. Use function notation.

Vocabulary.
Use the choices to complete each statement.

| **Domain** | **Function** | **Horizontal** |
| **Range** | **Relation** | **Vertical** |

1. Every linear equation is a function except for the equation that represents a _____

   line.

2. A _____ is the set of ordered pairs.

3. A _____ is a set of ordered pairs, that for every x-value there is only one y-

   value.

4. The _____ of a relation is the set of all its x-coordinates.

5. The _____ of a relation is the set of al its y-coordinates.

Objective 1

Find the domain and range of each relation.

6. $\{(1,3)(-3,8)(0,-5)(3,-4)\}$

6. _____

7. $\{(2,-7)(8,-1)(-7,0)(1,-2)\}$

7. _____

8. $\{(22,-33)(-13,19)(10,36)(-18,-17)\}$

8. _____

Objective 2

Determine if the given relation is a function.

9. $\{(2,-4)(-3,18)(10,18)(3,4)\}$

9. _____

Martin-Gay *Beginning and Intermediate Algebra, Fourth Edition*

10. $\{(1,-2)(4,4)(1,-5)(2,-3)\}$

10. _____

Determine whether the given graph is a function.

11.

11. _____

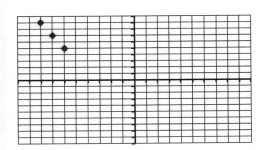

12.

12. _____

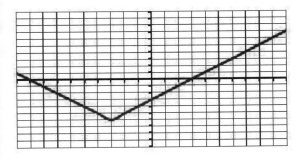

13.

13. _____

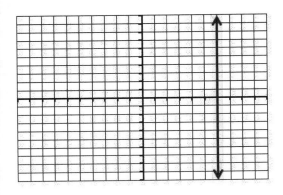

Determine whether the equation describes a function.

14. $y = 4x + 2$

14. _____

15. $\frac{1}{3}x - 1.7y = 0$ 15. _____

16. $x = y^2 - 7$ 16. _____

17. $x^2 - y = 21$ 17. _____

Objective 4

 18. Find $f(-2)$, $f(0)$, and, $f(3)$ for the given function. $f(x) = x^2 + 2$

                                                          18. _____

19. Find $g(-1)$, $g(2)$, and, $f(0)$ for the given function. Write each solution as an ordered pair.
    $g(x) = x^3 + 2x^2 - 3x + 6$

                                                          19. _____

Find the domain and range of each relation graphed.

20.   20. _____

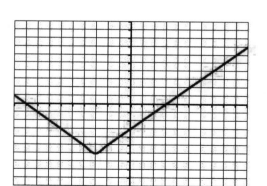

21.   21. _____

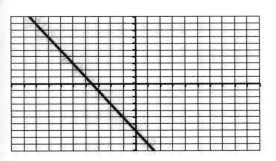

Find the domain of the given function.

22. $f(x) = x + 7$ 22. _____

23. $h(x) = \dfrac{4}{3 - 2x}$ 23. _____

24. $g(x) = |2x - 3|$ 24. _____

Concept Extension

25. Forensic scientists use the function $H(x) = 2.59x + 47.24$ to estimate the height of a woman in centimeters given the length x of her femur bone.

    a. Estimate the height of a woman whose femur measures 46 cm.

                         25 a. _____

    b. Estimate the height of a woman whose femur measures 39 cm.

                         25 b. _____

Chapter 3 Vocabulary

Vocabulary Word	Definition	Example
Rectangular coordinate system	A plane that contains a vertical (y) and horizontal (x) axes. The intersection of the axes is called the origin.	
Solution	An ordered pair the makes the equation true.	$y = 3x + 2$ Solution $(-2, -4)$ $-4 = 3(-2) + 2$
Linear Equation in Two Variable	An equation that can be written in standard form: $Ax + By = C$	$5x - 4y = 12$ $y = 2x - 8$
Intercept	A point on the graph where it hits a specific axis.	$y = 2x + 8$ Hits the y axis at 8.
Slope (m)	Rate of change. Steepness of the graph.	$\dfrac{y_2 - y_1}{x_2 - x_1}$
Slope-Intercept Form.	$y = mx + b$ where m is the slope and b is the y-intercept.	$y = 4x - 9$
Point-Slope Form	$y - y_1 = m(x - x_1)$ where m is the slope and the point (x_1, y_1)	Given m = 2 and $(1, 3)$ then $y - 3 = 2(x - 1)$
Relation	Set of ordered pairs. A function is a special relation that the x values do not repeat. For every x-value there is only one y-value.	$\{(1,2)(2,3)(4,5)\}$ Is a function as well.
Domain	Set of all the x-values in a relation.	For the relation above: D={1, 2, 4}
Range	Set of all the y-values in a relation.	For the relation above: R= {2, 3, 5}
$f(x)$	Function notation. A function of x.	$f(x) = 3x + 2$

Name:

Instructor:

Chapter 3 Practice Test A

Complete the ordered pair so that it is a solution to the given linear equation.

1. $y = -3x + 4$ $(-3,\ )(\ ,4)(7,\ )$

1. _____

State in which quadrant, if any, the given point lies.

2. $(-7, -8)$

2. _____

Determine whether the ordered pair is a solution of the given linear equation.

3. $4x - 3y = 8$ $(5, -4)$

3. _____

Determine whether the equation is a linear equation in two variables.

4. $x^2 = y - 19$

4. _____

Graph the given linear equation.

5. $4y = 2x - 8$

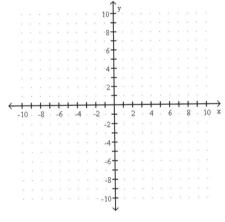

6. $y = -2x + 6$

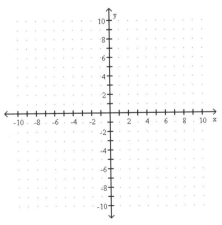

Identify the intercepts.

7.

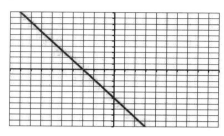

7.

Graph the linear equation by finding the x- and y-intercepts.

8. $6y - 9x = 18$

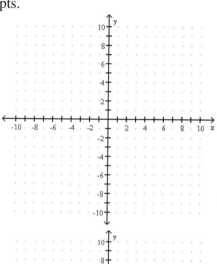

9. $x = -\dfrac{1}{2}y$

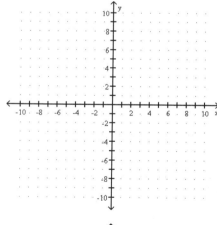

Graph the linear equation.

10. $x = -4$

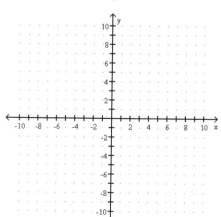

Find the slope of the line that goes through the given points.

11. $(-3,5)$ and $(5,7)$

11. _____

Find the slope of the line.

12.

12. _____

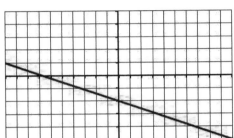

13. $3x - 7y = 9$

13. _____

Determine whether the given pair of lines are parallel, perpendicular, or neither.

14.

$y = \dfrac{4}{5}x - 9$

$5x - 4y = 12$

14. _____

Determine the slope and y-intercept of the given linear equation.

15. $10x - 18y = 24$

15. _____

Write the equation of the line in slope-intercept form.

16. $m = \dfrac{4}{7}$ \qquad $(0,-6)$

16. _____

Use the slope-intercept form the graph the equation.

17. $y = \dfrac{1}{3}x - 5$

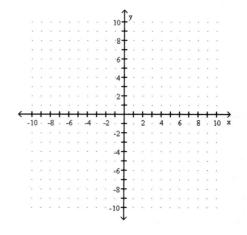

18. $5y - 4x = 15$

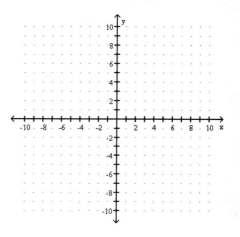

Write the equation of each line in standard form.

19. $m = -4$ $(-5, -7)$

19. _____

20. $m = \dfrac{2}{5}$ $(-3, 4)$

20. _____

21. A vertical line, through $(-14, 16)$.

21. _____

22. Through $(-33, 24)$, perpendicular to $x = 22$.

22. _____

Determine if the following is a function.

23. $\{(-1,5)(2,5)(-5,-5)\}$

23. _____

24. $y = |x - 4|$

24. _____

25. Given $h(x) = x^2 - 3x + 9$, find the indicated function values. Write the solutions as ordered pairs.

   $h(-4)$ $h(0)$ $h(3)$

25. _____

Chapter 3 Practice Test B

Complete the ordered pair so that it is a solution to the given linear equation.

1. $y = 5x - 8$ $\quad (-4,\ )(\ ,12)(2,\ )$ $\qquad\qquad\qquad$ 1. _____

    a. $(-4,28)(4,12)(2,2)$ $\qquad\qquad$ b. $(-4,-28)(4,12)(2,-2)$

    c. $(-4,28)(4,12)(2,-2)$ $\qquad\qquad$ d. $(-4,-28)(4,12)(2,2)$

State in which quadrant, if any, the given point lies.

2. $(3,-9)$ $\qquad\qquad\qquad\qquad\qquad\qquad\qquad$ 2. _____

    a. I $\qquad\qquad\qquad\qquad\qquad$ b. II

    c. III $\qquad\qquad\qquad\qquad\qquad$ d. IV

Determine whether the ordered pair is a solution of the given linear equation.

3. $-3x - 9y = 6$ $\quad (4,-2)$ $\qquad\qquad\qquad$ 3. _____

    a. Yes $\qquad\qquad\qquad\qquad\qquad$ b. No

Determine whether the equation is a linear equation in two variables.

4. $5.7y - 4.9x = 13.7$ $\qquad\qquad\qquad\qquad$ 4. _____

    a. Yes $\qquad\qquad\qquad\qquad\qquad$ b. No

Graph the given line.

5. $2x + 5y = 15$

5. _____

   a..

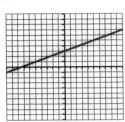

   b.

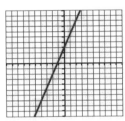

   c.

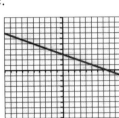

   d.

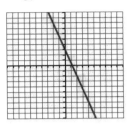

6. $y = -\dfrac{2}{3}x - 4$

6. _____

   a.

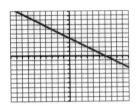

   b.

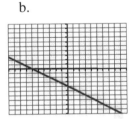

   c.

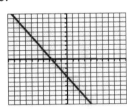

   d.

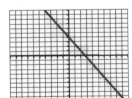

Determine whether the given pair of lines are parallel, perpendicular, or neither.

7.

   $3y = 6x - 4$

   $y = \dfrac{1}{2}x - 9$

7. _____

   a. Parallel

   b. Perpendicular

   c. Neither

   d. Identify the intercepts.

8.

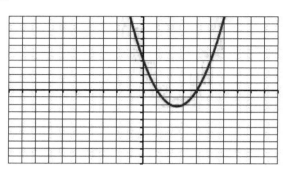

8. _____

    a. $(0,2)(0,3)(4,0)$ b. $(2,3)(4,0)$

    c. $(2,0)(3,0)(4,0)$ d. $(1,0)(3,0)(0,4)$

Find the slope of the line that goes through the given points.

9. $(-1,6)$ and $(3,-4)$

9. _____

    a. $\dfrac{2}{5}$ b. $-\dfrac{5}{2}$

    c. $\dfrac{5}{2}$ d. $-\dfrac{2}{5}$

Martin-Gay *Beginning and Intermediate Algebra, Fourth Edition*

Graph the linear equation by finding the x- and y-intercepts.

10. $9y - 4x = 18$

10. _____

    a.   b.

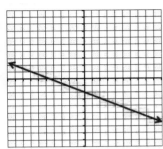   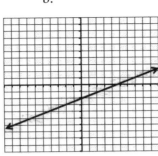

    c.   d

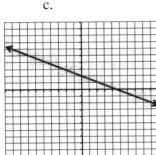   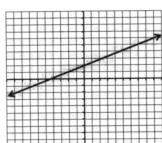

Write the equation of the line in slope-intercept form.

11. $m = \dfrac{3}{4}$ $(0,8)$

11. _____

    a. $y = \dfrac{3}{4}x + 8$ b. $y = \dfrac{3}{4}x - 8$

    c. $y = 8x + \dfrac{3}{4}$ d. $4x - 3y = 8$

Graph the linear equation.

12 $y = -3$

12. _____

    a.   b.

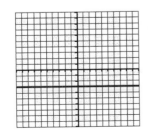   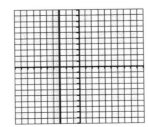

Graph the linear equation by finding the x- and y-intercepts.

13. $x = -4y$ 13. _____

a.

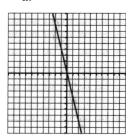

b.

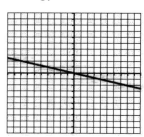

c.

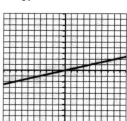

d..

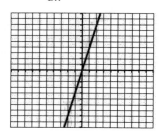

Find the slope of the line.

14.   14. _____

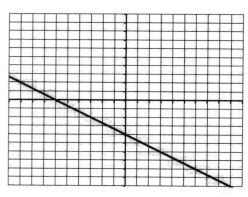

a. $\dfrac{2}{3}$ b. $\dfrac{3}{2}$

c. $-\dfrac{3}{2}$ d. $-\dfrac{2}{3}$

15. $-3x - 5y = 15$ 15. _____

    a. 3

    b. $\dfrac{5}{3}$

    c. $-\dfrac{3}{5}$

    d. $-\dfrac{5}{3}$

Determine the slope and y-intercept of the given linear equation.

16. $4x - 5y = 25$ 16. _____

    a. $m = \dfrac{5}{4} \quad b = 5$

    b. $m = \dfrac{4}{5} \quad b = -5$

    c. $m = -\dfrac{4}{5} \quad b = -5$

    d. $m = \dfrac{5}{4} \quad b = -5$

Write the equation of each line in standard form.

17. $m = 2 \quad (1,3)$ 17. _____

    a. $2x - y = 1$

    b. $2x - y = -1$

    c. $y = 2x - 1$

    d. $-2x - y = 1$

18. A horizontal line, through $(-8, 11)$ 18. _____

    a. $y = 11$

    b. $x = 11$

    c. $y = -8$

    d. $x = -8$

Use the slope-intercept form to graph the equation.

19. $y = -\frac{2}{5}x - 8$ 19.

    a.   b.

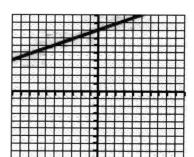   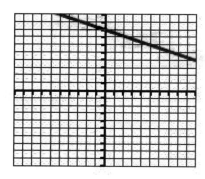

    c.   d.

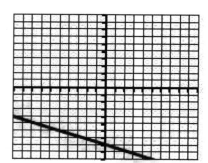   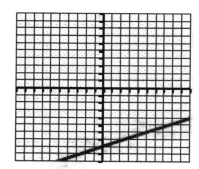

Determine if the following is a function.

20. $\{(2,3)(9,-3)(2,13)\}$ 20. _____

    a. It is a function.   b. It is not a function.

21. $y = 4$ 21. _____

    a. It is a function.   b. It is not a function.

22. Given $g(x) = |3x - 2|$, find the indicated function values. Write the solutions as ordered pairs.
  $g(-2)$ $g(0)$ $g(5)$

22. _____

  a. $(-2, -8)(0, -2)(5, 13)$

  b. $(-2, 8)(0, 2)(5, 13)$

  c. $(8, -2)(2, 0)(5, 13)$

  d. $(-2, -8)(0, -2)(5, 13)$

Use the slope-intercept form to graph the equation.

23. $7x + 4y = 28$

23. _____

  a.

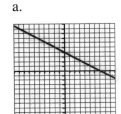

  b.

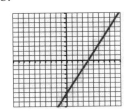

  c.

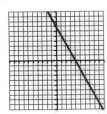

  d.

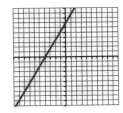

Write the equation of each line in standard form.

24. Through $(13, -14)$, parallel to $x = 2$.

24. _____

  a. $x = 13$

  b. $y = 13$

  c. $x = -14$

  d. $y = -14$

25. $m = -\dfrac{1}{3}$ $(-6,1)$

25. _____

    a. $x + 3y = -3$ b. $3x - y = 3$

    c. $-x + 3y = 9$ d. $x + 3y = 9$

Chapter 4 Solving Systems of Linear Equations
Section 4.1 Solving Systems of Linear Equations by Graphing

Learning Objectives
1. Determine if an ordered pair is a solution of a system of equations in two variables.
2. Solve a system of linear equations by graphing.
3. Without graphing, determine the number of solutions of a system.

Vocabulary.
Use the choices to complete each statement.

Consistent	**Dependent**	**Inconsistent**
Independent	**Solution**	**System of linear equations**

1. If the system of equations has two distinct lines, it is called a(n) _____ system.

2. A set of two or more linear equations is called a(n) _____.

3. If a system of equations has at least one or more solutions, it is called a(n) _____

   system.

4. When graphing a system of equations, if both lines turn out to be the same line, it is called a(n)

   _____ system.

5. Parallel lines make a(n) _____ system because the system will have no

   solution.

6. A _____ of a system of equations is an ordered pair that not only works in one of the

   equations, but all the equations in the system.

Objective 1
Determine whether each ordered pair is a solution of the system of linear equations.

7. $\begin{cases} x + y = 16 \\ y - x = 2 \end{cases}$

     a. $(7, 9)$ 7 a. _____

     b. $(10, 6)$ 7b. _____

8. $\begin{cases} 3x - y = 5 \\ x + 2y = 11 \end{cases}$

     a. $(3, 4)$ 8a. _____

     b. $(0, -5)$ 8b. _____

9. $\begin{cases} 2y = -3x - 8 \\ 5x - 6y = -4 \end{cases}$

    a. $(4, 4)$

9a. _____

    b. $(-2, -1)$

9b. _____

Objective 2

Solve the system by graphing.

10. $\begin{cases} x = \dfrac{3}{4}y + 1 \\ 7x - 3y = -2 \end{cases}$

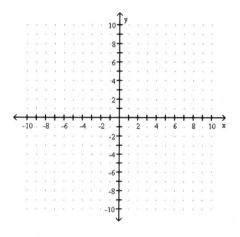

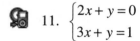 11. $\begin{cases} 2x + y = 0 \\ 3x + y = 1 \end{cases}$

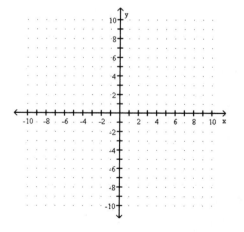

12. $\begin{cases} 2x - 4y = -6 \\ 3x - 6y = 10 \end{cases}$

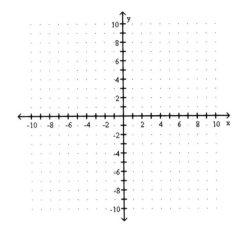

13. $\begin{cases} y = -2x + 3 \\ 6x + 3y = 9 \end{cases}$

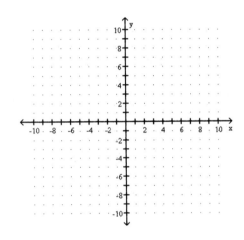

Objective 3

Without graphing, decide if the graphs of the equations are identical lines, parallel lines, or lines that are intersecting. Also state how many solutions will the system have?

14. $\begin{cases} 4x + y = 24 \\ x + 2y = 2 \end{cases}$

14. _____

15. $\begin{cases} y = \dfrac{2}{3}x - \dfrac{4}{3} \\ 6x = 12 + 9y \end{cases}$

15. _____

16. $\begin{cases} 3x - 7y = -9 \\ 3x - 7y = 5 \end{cases}$

16. _____

17. $\begin{cases} \dfrac{2}{3}x + \dfrac{1}{5}y = 0 \\ \dfrac{3}{2}x - \dfrac{3}{10}y = -15 \end{cases}$

17. _____

Concept Extensions

18. Is $\left(\dfrac{37}{13}, -\dfrac{21}{13} \right)$ a solution of the system: $\begin{cases} 5x + 2y = 11 \\ 7x + 8y = 7 \end{cases}$?

18. _____

Section 4.2 Solving Systems of Linear Equations by Substitution

Learning Objectives
 1. Use the substitution method to solve a system of linear equations.

Objective 1

Solve each system of equations by the substitution method.

1. $\begin{cases} x + y = 6 \\ y = -3x \end{cases}$

1. _____

2. $\begin{cases} 3x - y = 10 \\ y = -x + 2 \end{cases}$

2. _____

3. $\begin{cases} -x + 3y = -3 \\ 9y - 3x = -9 \end{cases}$

3. _____

4. $\begin{cases} y = -\dfrac{3}{4}x - 7 \\ y = -\dfrac{1}{2}x - 3 \end{cases}$

4. _____

5. $\begin{cases} -8x + 3y = 22 \\ 4x + 3y = -2 \end{cases}$

5. _____

 6. $\begin{cases} 3x + 6y = 9 \\ 4x + 8y = 16 \end{cases}$

6. _____

7. $\begin{cases} 2x - 3y = -4 \\ 3x + 4y = -6 \end{cases}$

7. _____

8. $\begin{cases} 4x - 2y = 5 \\ x = \dfrac{1}{2}y + \dfrac{5}{4} \end{cases}$

8. _____

9. $\begin{cases} \dfrac{1}{3}x - y = 2 \\ x - 3y = 6 \end{cases}$

9. _____

10. $\begin{cases} 2x + y = 9 \\ 7x + 4y = 1 \end{cases}$

10. _____

11. $\begin{cases} -2y = 1 + 7x \\ 4x + 5y = 2 \end{cases}$

11. _____

12. $\begin{cases} 2x + 3y = 6 \\ y = 3x - 5 \end{cases}$

12. _____

13. $\begin{cases} 3x - 6y = 10 \\ 2x - 4y = -6 \end{cases}$

13. _____

14. $\begin{cases} y = 22 - x \\ 0.05x + 0.10y = 1.70 \end{cases}$

14. _____

15. $\begin{cases} 2x + 6y = -28 \\ 3x + 9y = -27 \end{cases}$

15. _____

16. $\begin{cases} 7x - 6y = -1 \\ x = 2y - 1 \end{cases}$

16. _____

17. $\begin{cases} 4x + 5y = 2 \\ 16x - 15y = 1 \end{cases}$

17. _____

Concept Extensions

18.

$$\begin{cases} \dfrac{x+2}{3} = \dfrac{3-y}{2} \\ \dfrac{x+3}{2} = \dfrac{2-y}{3} \end{cases}$$

18. _____

19. $\begin{cases} 2(x-1) - 3(y+2) = 30 \\ 3(x+2) + 2(y-1) = -4 \end{cases}$

19. _____

Martin-Gay *Beginning and Intermediate Algebra, Fourth Edition*

Section 4.3 Solving Systems of Linear Equations by Addition

Learning Objectives
   1. Use the addition method to solve a system of linear equations

Objective 1
Solve each system of equations by the addition method.

1. $\begin{cases} x - 2y = 8 \\ -x + 5y = -17 \end{cases}$

1. _____

2. $\begin{cases} 2x + 9y = 2 \\ x - 2y = -12 \end{cases}$

2. _____

3. $\begin{cases} 5x + 2y = -4 \\ 5x - 3y = 6 \end{cases}$

3. _____

4. $\begin{cases} x + y = 48 \\ 12x + 14y = 628 \end{cases}$

4. _____

5. $\begin{cases} 10x - 2y = 2 \\ 5x - y = 1 \end{cases}$

5. _____

6. $\begin{cases} x + 2y = 0 \\ 2x - y = 0 \end{cases}$

6. _____

7. $\begin{cases} 2x = 3(y - 2) \\ 2(x + 4) = 3y \end{cases}$

7. _____

8. $\begin{cases} 8x - 4y = 18 \\ 3x - 8 = 2y \end{cases}$

8. _____

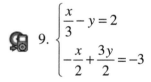 9. $\begin{cases} \dfrac{x}{3} - y = 2 \\ -\dfrac{x}{2} + \dfrac{3y}{2} = -3 \end{cases}$

9. _____

10. $\begin{cases} \dfrac{1}{2}x - \dfrac{1}{4}y = 1 \\ \dfrac{1}{3}x + y = 3 \end{cases}$

10. _____

11. $\begin{cases} \dfrac{7}{12}x - \dfrac{1}{2}y = \dfrac{1}{6} \\ \dfrac{2}{5}x - \dfrac{1}{3}y = \dfrac{11}{15} \end{cases}$

11. _____

12. $\begin{cases} 3.5x + 2.5y = 17 \\ -1.5x - 7.5y = -33 \end{cases}$

12. _____

13. $\begin{cases} 3x - 2.5y = 0 \\ 2.5x - 3y = -1 \end{cases}$

13. _____

14. $\begin{cases} 0.1x + 0.06y = 9 \\ 0.09x + 0.5y = 52.7 \end{cases}$

14. _____

15. $\begin{cases} 3x + 2y = 3 \\ 4x - 3y = -13 \end{cases}$

15. _____

16. $\begin{cases} 0.07x + 0.3y = 6.70 \\ 7x + 30 = 67 \end{cases}$

16. _____

17. $\begin{cases} 4x + 7y = 2 \\ 9x - 2y = 1 \end{cases}$

17. _____

Concept Extension

18. $\begin{cases} \dfrac{x-y}{2} - \dfrac{2x-y}{3} = -\dfrac{1}{4} \\ \dfrac{2x+y}{3} + \dfrac{x+y}{2} = \dfrac{17}{6} \end{cases}$

18. _____

19. If possible find a solution to this system: $\begin{cases} x + y = 5 \\ x - y = -3 \\ 2x - y = -2 \end{cases}$

19. _____

Section 4.4 Solving Systems of Linear Equations in Three Variables

Learning Objectives

    1. Solve a system of three linear equations in three variables.

Objective 1

Solve each system.

1. $\begin{cases} x + y - z = 4 \\ 3x + 2y - 3z = 6 \\ x - y + 4z = 8 \end{cases}$

                                                1. _____

2. $\begin{cases} x + 2y = 4 \\ x + z = -3 \\ 3x + y - z = 5 \end{cases}$

                                                2. _____

3. $\begin{cases} x + 2y - z = 4 \\ 3x - y + 2z = 6 \\ x + y = 5 \end{cases}$

                                                3. _____

4. $\begin{cases} x - y + 2z = 11 \\ 2x - 4y + z = 4 \\ x + 3y - z = 6 \end{cases}$

4. _____

5. $\begin{cases} 2x - 3y + z = 2 \\ x - 5y + 5z = 3 \\ 3x + y - 3z = 5 \end{cases}$

5. _____

6. $\begin{cases} x + 3y - 4z = 6 \\ 3x + 9y - 12x = 18 \\ -5x - 15y + 20z = -30 \end{cases}$

6. _____

7. $\begin{cases} x - z = 0 \\ 2x + y + z = 7 \\ 3y - 2z = 10 \end{cases}$

7. _____

8. $\begin{cases} 8x - 2y + 6z = 28 \\ 3x + y - 2z = 9 \\ 2x - 6y + 3z = -5 \end{cases}$

8. _____

9. $\begin{cases} 4x + 4y - 3z = -22 \\ 5x - 6y - 4z = 15 \\ 3x + 5y + z = -5 \end{cases}$

9. _____

10. $\begin{cases} 2x + 4y + 3z = -9 \\ 3x + 2y + 5z = -19 \\ 4x - 5y + 2z = -19 \end{cases}$

10. _____

Concept Extension

11.

$\begin{cases} 2x + 4y + 2z + 3w = -7 \\ -3x - 2y + 5z + 4w = 9 \\ 4x - 3y - 2z - 3w = 11 \\ -2x + 3y - 5z - 4w = -16 \end{cases}$

11. _____

Martin-Gay *Beginning and Intermediate Algebra, Fourth Edition*

Section 4.5 Systems of Linear Equations and Problem Solving

Learning Objectives

1. Solving problems that can be modeled by a system of two linear equations.

2. Solve problems with cost and revenue functions.

3. Solve problems that can be modeled by a system of three linear equations.

Objective 1

Solve.

1. One number is five more than twice the second number. If the difference between twice the first number and three times the second number is six, what are the two numbers?

1. _____

2. Find how many quarts of 4% butterfat milk and 1% butterfat milk should be mixed to yield 60 quarts of 2% butterfat milk.

2. _____

3. Ann Marie Jones has been pricing Amtrak train fares for a group trip to New York. Three adults and four children must pay $159. Two adults and three children must pay $112. Find the price of an adult's ticket, and find the price of a child's ticket.

3. _____

4. The perimeter of a rectangle is 52 feet. The length of the rectangle is 6 more than triple the width. Find the dimensions of the rectangle.

4. _____

5. Two joggers leave an apartment building and jog in opposite directions. One jogger travels at 2 mph faster than the other jogger. If after 2.5 hours, they are 15 miles apart, what is the rate of each jogger?

5. _____

Objective 2

Given the cost function and the revenue function, find the number of units x that must be sold to break even.

6. $C(x) = 1.7x + 1700$ $R(x) = 2.4x$

6. _____

7. $C(x) = 19x + 2250$ $R(x) = 34x$

7. _____

8. The planning department of Abstract Office Supplies has been asked to determine whether the company should introduce a new computer desk next year. The department estimates that $6000 of new manufacturing equipment will need to be purchased and that the cost of constructing each desk will be $200. The department also estimates that the revenue from each desk will be $450.

a. Determine the revenue function $R(x)$ from the sale of x desks.

8a. _____

b. Determine the cost function $C(x)$ for manufacturing x desks.

8b. _____

c. Find the break-even point.

8c. _____

9. The U-Haul-It company is looking to purchase a new $20,000 truck for its fleet. The company rents out the trucks for $100 each rental. When the truck is returned, it will cost U-Haul-it $50 to clean and get the truck ready for its next renter. When will the company finally start earning money for this vehicle.

9. _____

Objective 3

10. One number is four more than the second number. Triple the first number is 1 more than the third number. The sum of all three numbers is 23. Find the numbers.

10. _____

11. The sum of the angles of a triangle is 180 degrees. The sum of the twice the smallest angle and the largest angle is 175 degrees. Three times the second angle less the largest angle is 65 degrees. Find the measures of the angles of the triangle.

11. _____

Concept Extension

12. Find the values of a, b, and c such that the equation $y = ax^2 + bx + c$ has the ordered pair solutions of $(3, 27)$, $(-5, 131)$, and $(-2, 32)$.

12. _____

Chapter 4 Vocabulary

Vocabulary Word	Definition	Example
System of Linear Equations	A set of two or more equations.	$\begin{cases} x+y=7 \\ 2x+5y=13 \end{cases}$
Solution of a System	An ordered pair (or triple) that satisfies all equations of that system.	$\begin{cases} x+y=7 \\ x-y=-1 \end{cases}$ $(3,4)$ is a solution of the given system.
Consistent System	A system that has at least one solution.	$\begin{cases} x+y=5 \\ x-y=6 \end{cases}$ is consistent
Inconsistent System	A system that does not have a solution.	$\begin{cases} x+2y=6 \\ 4x+8x=10 \end{cases}$ is inconsistent
Dependent System	A system that all equations of the system are identical lines.	$\begin{cases} x+y=2 \\ 4x+4y=8 \end{cases}$ is dependent
Independent System	A system where each equation will result in a distinct line.	$\begin{cases} 4x+5y=7 \\ 2x+3y=4 \end{cases}$ is independent

Practice Test A

1. Is $(-2,-3)$ a solution of $\begin{cases} y = x + 1 \\ 3x = 2y \end{cases}$

1. _____

2. Is $(1,5)$ a solution of $\begin{cases} 2x + y = 7 \\ 3x - 4y = 17 \end{cases}$

2. _____

Solve the following system of equations by graphing.

3. $\begin{cases} x + y = 5 \\ x + 2y = 4 \end{cases}$

3. _____

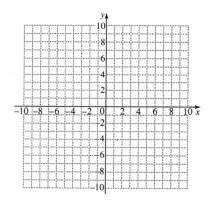

Solve the system using substitution.

4. $\begin{cases} 4x - 3y = 29 \\ 2x + y = 7 \end{cases}$

4. _____

Martin-Gay *Beginning and Intermediate Algebra, Fourth Edition*

Solve the system using elimination.

5. $\begin{cases} -3x+5y=-13 \\ 4x-2y=-6 \end{cases}$

5. _____

Solve the following systems using the method of your choice.

6. $\begin{cases} x+3y=4 \\ x-2y=-11 \end{cases}$

6. _____

7. $\begin{cases} 2x+4y+z=4 \\ -3x-2y+3z=-17 \\ 4x-y-4z=8 \end{cases}$

7. _____

8. $\begin{cases} 10x-12y=6 \\ x=\frac{6}{5}y+\frac{3}{5} \end{cases}$

8. _____

9. $\begin{cases} 5x + 3y = -1 \\ -3x - 4y = -6 \end{cases}$

9. _____

10. $\begin{cases} x + 3y - z = 0 \\ 2x - 4y + z = -10 \\ -3x + 5y - 4z = 2 \end{cases}$

10. _____

11. $\begin{cases} x + 2y = 6 \\ 4x - 8y = 9 \end{cases}$

11. _____

Solve.

12. One number is 6 more than half the second number. The sum of the three times the first number and twice the second number is negative three.

12. _____

13. The sum of a three digit number is 15. The third digit is one more than twice the first digit. If you double the third digit the result is the same as the sum of the first two digits. What is the three digit number?

13. _____

14. A college was selling tickets to its annual fine arts department art show. The price for adults was $8 and the price for children was $5. If the department made $572 dollars, and a total of 82 tickets were sold, how many on each type of ticket were sold?

14. _____

15. Given that $C(x) = 0.90x + 24,000$ and $R(x) = 2.10x$, find the break –even point.

15. _____

Name:

Instructor:

Practice Test B

Solve the system by substitution.

1. $\begin{cases} 2x+6y=8 \\ 3x-y=12 \end{cases}$

1. _____

    a. $(1,2)$ b. $(5,3)$

    c. $(4,0)$ d. $(2,-2)$

2. $\begin{cases} 5x=-3y+3 \\ 3y=4x-6 \end{cases}$

2. _____

    a. $(0,1)$ b. $\left(1,\frac{2}{3}\right)$

    c. $\varnothing$ d. $\{(x,y)\,|\,5x=-3y+3\}$

Solve the system using elimination.

3. $\begin{cases} \dfrac{1}{3}x-\dfrac{3}{4}y=-\dfrac{4}{3} \\ \dfrac{1}{2}x+\dfrac{3}{8}y=4 \end{cases}$

3. _____

    a. $\left(\frac{1}{2},2\right)$ b. $(5,4)$

    c. $\varnothing$ d. $\left\{(x,y)\,|\,\dfrac{1}{3}x-\dfrac{3}{4}y=-\dfrac{4}{3}\right\}$ $(2,-2)$

4. $\begin{cases} 2x - y = -1 \\ 8x + 10y = -18 \end{cases}$

4. _____

    a. $(-1, -1)$ b. $(-2, -3)$

    c. $\varnothing$ d. $\{(x, y) \mid 2x - y = -1\}$

Solve the system of equations using the method of your choice.

5. $\begin{cases} 2x - 3y + z = 18 \\ 3x + 4y + 2z = -4 \\ -2x - 5y - 3z = 4 \end{cases}$

5. _____

    a. $(0, 1)$ b. $(-2, 3, -5)$

    c. $\varnothing$ d. $\{(x, y, z) \mid 2x - 3y + z = 18\}$

6. $\begin{cases} 3x - 6y = 18 \\ -\frac{3}{2}x + 2y = -6 \end{cases}$

6. _____

    a. $(0, 3)$ b. $(6, 0)$

    c. $\varnothing$ d. $\{(x, y) \mid 3x - 6y = 18\}$

Martin-Gay *Beginning and Intermediate Algebra, Fourth Edition*

7. $\begin{cases} x + 2y + 3z = 7 \\ -2x - 3y + z = 24 \\ 3x + 4y + 2z = -13 \end{cases}$

7. _____

    a. $(0, 2, 1)$

    b. $(-3, -4, 6)$

    c. $\varnothing$

    d. $\{(x, y, z) \mid x + 2y + 3z = 7\}$

8. $\begin{cases} 3x + 4y = 8 \\ y = -\frac{3}{4}x - 2 \end{cases}$

8. _____

    a. $(0, 2)$

    b. $(-4, 1)$

    c. $\varnothing$

    d. $\{(x, y) \mid 3x + 4y = 8\}$

Solve the system by graphing.

9. $\begin{cases} x + 2y = 6 \\ -3x + 4y = -8 \end{cases}$

9. _____

    a. $(4, 1)$

    b. $(0, 3)$

    c. $\varnothing$

    d. $\{(x, y) \mid x + 2y = 6\}$

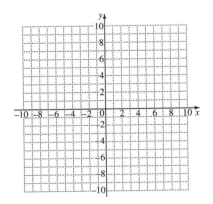

Solve.

10. Joe has 13 coins in his piggy bank. He only has three types of coins quarters, dimes, and nickels. If Joe doubles the number of nickels, he will have 2 more nickels then the total of dimes and quarters. Find how many of each coin Joe has if he has $1.35 in his piggy bank.

                                                                          10. _____

   a. 3 quarters, 4 dimes, 6 nickels   b. 4 quarters, 5 dimes, 4 nickels

   c. 3 quarters, 7 dimes, 3 nickels   d. 2 quarters, 6 dimes, 4 nickels

11. The perimeter of a triangle is 24 inches. The first side is 3 more than the second side. The sum of the second and third side is two more than the first side. Find the three sides.

                                                                          11. _____

   a. 10, 7, and 7   b. 9, 6, and 9

   c. 5, 8, and 11   d. 12, 9, and 3

12. The sum of a number and three times a number is four. The sum of three times the first number and nine times the second number is twelve. Find the numbers.

                                                                          12. _____

   a. 1 and 1   b. -2 and 2

   c. $\varnothing$ d. All real numbers

13. Find the measure of the angles of a triangle if the measure of one angle is twice the measure of the second angle and the third angle doubled is the same as the sum of the other two angles. Remember that the sum on the angles in a triangle total 180 degrees.

13. _____

   a. $40°, 65°, 75°$ b. $30°, 70°, 80°$

   c. $40°, 60°, 80°$ d. $50°, 60°, 70°$

14. Two cars left Chicago, IL traveling in opposite directions. After 4 hours they were 464 miles apart. If one car travels at 4 mph faster than the other car, find the rate of each car.

14. _____

   a. 52 mph and 56 mph   b. 58 mph and 62 mph

   c. 56 mph and 60 mph   d. 60 mph and 64 mph

15. Given the cost function $C(x) = 15x + 25000$ and the revenue function, $R(x) = 25x$. Find the break even point.

15. _____

   a. 2000 units   b. 2500 units

   c. 625 units   d. 250 units

Chapter 5 Exponents and Polynomials
5.1 Exponents

Learning Objectives
1. Evaluate exponential expressions.
2. Use the product rule for exponents.
3. Use the power rule for exponents.
4. Use the power rules for products and quotients.
5. Use the quotient rule for exponents and define a number raised to the 0 power.
6. Decide which rule(s) to use to simplify an expression.

Vocabulary.
Use the choices to complete each statement.

Add	Base	Divide	Exponents
Exponent	False	Multiply	Subtract
True	0	1	

1. In the expression 6^9, six is the _____ and nine is the _____.

2. According to the quotient rule, to simplify $x^{10} \div x^3$, we will _____ the exponents.

3. If given the expression $\left(x^5\right)^3$, to simplify we would _____ the exponents.

4. True or False. $(-3)^0 = -1$. _____

5. If there is no exponent, i.e. x, the exponent is understood to be _____.

Objective 1
Evaluate each expression.

6. $\left(\dfrac{3}{4}\right)^4$

6. _____

7. $(0.03)^3$

7. _____

8. -16^2

8. _____

Objective 2
Use the product rule to simplify. Write the results using exponents.

9. $\left(2^3 x^5\right)\left(2^8 x^{13}\right)$

9. _____

10. $m^{27}n^8\left(m^{10}n^{21}\right)$

10. _____

11. $\left(x^9y\right)\left(x^{10}y^5\right)$

11. _____

Objective 3
Use the power rule to simplify. Write the results using exponents.

12. $\left(5^3\right)^8$

12. _____

13. $\left[(-7)^4\right]^{12}$

13. _____

14. $\left(x^4\right)^{18}$

14. _____

Objective 4
Use the power rule and power of the product rule to simplify each expression.

15. $\left(a^4b^5\right)^6$

15. _____

16. $\left[(-4)^3x^4\right]^9$

16. _____

17. $\left(0.2y^{16}\right)^3$

17. _____

Objective 5
Use the power rule and the power of a quotient to simplify each expression.

18. $\left(\dfrac{x}{6}\right)^4$

18. _____

 19. $\left(\dfrac{-2xz}{y^5}\right)^2$

19. _____

20. $\left(\dfrac{a^2 b^4}{3c^5}\right)^5$

20. _____

21. $(-8)^0$

21. _____

22. $-14x^0$

22. _____

23. $\dfrac{m^0}{9}$

23. _____

Objective 6
Simplify each expression.

 24. $\dfrac{9a^4 b^7}{27ab^2}$

24. _____

25. $\dfrac{\left(4m^3 n^5\right)^3}{\left(2m^4 n^6\right)^2}$

25. _____

Concept Extensions

26. Prove the fact that $7^0 = 1$. Hint: the quotient rule.

26. _____

27. The volume of a cube is given but he formula $V = x^3$, where x is the length of the side. How would the volume of a cube change if the side was quadrupled?

27. _____

Section 5.2 Polynomial Functions and Adding and Subtracting Polynomials

Learning Objectives
1. Define polynomial, monomial, binomial, trinomial, and degree.
2. Define polynomial functions.
3. Simplify a polynomial by combining like terms.
4. Add and subtract polynomials.

Vocabulary
Use the choices to complete each statement.

Binomial	**Coefficient**	**Constant**
False	**Monomial**	**True**

1. A two term polynomial is referred to as a(n) _____.
2. A numerical term in a polynomial without any variables is called a(n) _____.
3. True or false. The degree of the term is the greatest degree of the polynomial.
4. A single term polynomial is called a(n) _____.
5. The numerical _____ is the numerical factor of the term.

Objective 1

Find the degree of each of the following polynomials and determine whether it is a monomial, binomial, trinomial, or none of these.

6. $5x^2 + 7$

6. _____

 7. $12x^4y - x^2y^2 - 12x^2y^4$

7. _____

8. $3m^2n + 4mn^2 - 7m^2n^2 - mn$

8. _____

9. $\frac{1}{2}x^2y - \frac{1}{4}y^3 - \frac{1}{8}xy^4 + \frac{1}{3}x^5y^3$

9. _____

10. $3a^2c^3 - 2c$

10. _____

Objective 2

11. If $P(x) = x^2 + 2x + 1$, find $P(-1)$.

11. _____

12. If $Q(x) = 3x^4 - 4x^3 + 2x^2 - 5$, find $Q(-1)$.

12. _____

13. If $P(x) = 6x^2 - 7x + 9x^3$, find $P(-1)$.

13. _____

Objective 3
Simplify each of the following by combining like terms.

14. $3x^4 - 5x^3 + 7x - 6x^3 - x^4$

14. _____

15. $5ab - 6a + 10ba - 8b$

15. _____

Objective 4
Perform the indicated operations.

16. $(4x - 3) + (8x + 7)$

16. _____

17. $(7x - 5) + (-17x - 8)$

17. _____

18. $(2x^2 + 5) - (3x^2 - 9)$

18. _____

19. $\left(3x^3 + 4x - 1\right) + \left(5x^3 - 4x^2\right) - \left(7x^3 + 3x^2 - 4\right)$

19. _____

20.

$$5x^3 - 4x^2 + 6x - 2$$
$$\underline{-\left(3x^3 - 2x^2 - x - 4\right)}$$

20. _____

21. Subtract $\left(6m^2 - 7m + 1\right)$ from the sum of $\left(4m^2 - 3m\right)$ and $\left(8m - 9\right)$.

21. _____

22. $\left[\left(3x^2 - 5x + 4\right) - \left(5x^2 - 7x - 2\right)\right] + \left(6x^2 - 12x - 8\right)$

22. _____

Concept Extensions

23. An object is thrown from a cliff. The height of the object is given by the polynomial function $P(t) = -16t^2 - 32t + 496$; where t represents time. How high is the object after 4 seconds?

23. _____

Objective 5.3 Multiplying Polynomials

Learning Objectives
1. Use the distributive property to multiply polynomials.
2. Multiply polynomials vertically.

Objective 1
Multiply.

1. $2x^2\left(4x^3 - 3x\right)$

1. _____

2. $-\dfrac{1}{2}y\left(4y^4 - 8y^3 - 16y + 10\right)$

2. _____

3. $-y\left(4x^3 - 7x^2y + xy^2 + 3y^3\right)$

3. _____

4. $(x-7)(x+4)$

4. _____

5. $\left(\dfrac{1}{2}x - \dfrac{1}{3}\right)\left(\dfrac{1}{5}x - \dfrac{1}{4}\right)$

5. _____

6. $(m+3)(m-2)$

6. _____

7. $(a+7)(b-10)$

7. _____

8. $(x-7)^2$

8. _____

9. $(3x^2+1)^2$

9. _____

10. $(x+2)(x^2-3x-4)$

10. _____

11. $(x-2)^3$

11. _____

12. $(m^2-3m+1)(m^2+4m-2)$

12. _____

Objective 52
Multiply vertically.

13. $(a-4)(a+1)$

13. _____

14. $(3m+n)(4m-3n)$

14. _____

15. $(5x+1)(2x^2+4x-1)$

15. _____

16. $\left(y^2 - 3y + 4\right)\left(y^2 + 2y - 1\right)$

16. _____

Concept Extensions.

17. Write a polynomial that describes the area of the shaded region.

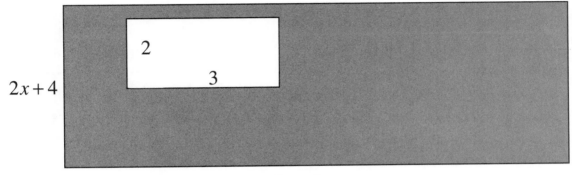

17. _____

Section 5.4 Special Products.

Learning Objectives
1. Multiply two binomials using the FOIL method.
2. Multiply the sum and difference of two terms.

Objective 5.4.1
Multiply using the FOIL method.

1. $(x-7)(x+6)$

1. _____

2. $(m+3)(m+9)$

2. _____

3. $(3y-4x)(11y-12x)$

3. _____

4. $(14a-9)(-a+7)$

4. _____

5. $(-2m-p)(-8m+p)$

5. _____

6. $(3b+7)(2b-15)$

6. _____

Objective 2
Multiply.

7. $\left(x-\dfrac{1}{4}\right)^2$

7. _____

8. $(5x+9)^2$

8. _____

9. $(7s - 3t)^2$

10. $\left(\dfrac{1}{3}x - \dfrac{1}{2}y\right)^2$

10. _____

Objective 3
Multiply.

11. $(x + 5)(x - 5)$

11. _____

12. $\left(\dfrac{1}{2}x - 1\right)\left(\dfrac{1}{2}x + 1\right)$

12. _____

13. $(9x + y)(9x - y)$

13. _____

14. $(7x - 1.5)(7x + 1.5)$

14. _____

15. $(9x - 13)(9x + 13)$

15. _____

Concept Extensions.

16. $\left[(2x - 1) + 3y\right]\left[(2x - 1) - 3y\right]$

16. _____

17. $(5x - 2)^4$

17. _____

Section 5.5 Negative Exponents and Scientific Notation

Learning Objectives
1. Simplify expressions containing negative exponents.
2. Use all the rules and definitions for exponents to simplify exponential expressions.
3. Write numbers in scientific notation.
4. Convert numbers from scientific notation to standard form.

Vocabulary
Use the choices to complete each statement.

Standard notation

$\dfrac{1}{x^7}$

Scientific notation

x^7

1. 3.2×10^{-4} is written in _____

2. The expression x^{-7} equals _____.

3. The expression $\dfrac{1}{x^{-7}}$ equals _____.

4. 0.000032 is written in _____.

Objective 1
Simplify each expression. Write each result using positive exponents.

5. 8^{-3}

5. _____

6. $\dfrac{1}{7^{-5}}$

6. _____

7. $-4x^{-6}$

7. _____

8. $\dfrac{m^{-3}}{n^{-4}}$

8. _____

9. $3^{-2} - 3^0$

9. _____

Objective 2
Simplify each expression. Write each result using positive exponents.

10. $\dfrac{r}{r^{-3}r^{-2}}$

10. _____

11. $\dfrac{\left(x^3\right)^{-4}}{x^{-8}}$

11. _____

12. $\dfrac{\left(2x^4y^5\right)^{-2}}{\left(4x^7y^3\right)^{-3}}$

12. _____

13. $\left(-4a^4b^3\right)\left(-2a^{-2}b^4\right)^2$

13. _____

Objective 5.5.3
Write each number in scientific notation.

14.   897,000,000

14. _____

15.   0.00000167

15. _____

16.   17, 000,000,000

16. _____

Objective 4
Write each number in standard notation.

17. 3.478×10^{-7}

17. _____

18. 8.7325×10^8

18. _____

Concept Extensions.

Evaluate each expression using exponential rules. Write each result in standard notation.

 19. $\dfrac{1.4 \times 10^{-2}}{7 \times 10^{-8}}$

19. _____

20. $\left(3.2 \times 10^{-3}\right)\left(6.4 \times 10^4\right)$

20. _____

Section 5.6 Dividing Polynomials.

Learning Objectives.
1. Divide a polynomial by a monomial.
2. Use long division to divide a polynomial by another polynomial.

Objective 1
Perform each division.

1. $\left(6x^3 - 5x^2 + 4x\right) \div (-2x)$

1. _____

2. Divide $\left(24a^4b^3 + 32a^4b^4 - 16a^2b^2\right)$ by $4a^2b$

2. _____

3. $\dfrac{-9x^5 + 3x^4 - 12}{3x^3}$

3. _____

4. $\dfrac{14m^2 - 8m - 6}{-4m}$

4. _____

Objective 2
Find each quotient using long division.

5. $\dfrac{x^2 + 4x + 3}{x + 3}$

5. _____

6. $\dfrac{12x^2 - 11x - 3}{4x - 1}$

6. _____

7. $\left(8m^3 + 24m^2 + 2m - 14\right) \div \left(2m + 3\right)$

7. _____

8. $\dfrac{x^3 - 1}{x - 1}$

8. _____

9. $\dfrac{2b^3 + 9b^2 + 6b - 4}{b + 4}$

9. _____

10. $\dfrac{7 - 11x^2}{x + 1}$

10. _____

11. $\dfrac{m^5 - m^2}{m^4 - m}$

Concept Extensions

Divide using long division.

12. $\dfrac{x^4 - x^3 - 4x^2 + 19x - 15}{x^2 - 3x + 5}$

Section 5.7 Synthetic Division and the Remainder Theorem

Learning Objectives
1. Use synthetic division to divide a polynomial by a binomial.
2. Use the remainder theorem to evaluate polynomials.

Objective 1

Use synthetic division to divide.

1. $\left(x^2 + 7x + 6\right) \div \left(x + 6\right)$

1. _____

2. $(x^3 - 7x^2 - 13x + 5) \div (x - 2)$

2. _____

3. $\dfrac{x^3 + 27}{x + 3}$

3. _____

4. $\left(2x^4 + \dfrac{1}{2}x^3 + x^2 + x\right) \div \left(x - 2\right)$

4. _____

5. $(9x^2 - 8) \div (x + 2)$

5. _____

6. $\dfrac{3x^4 + 8x^3 + 5x}{x - 1}$

6. _____

7. $(7x^2 - 4x + 12 + 3x^3) \div (x + 1)$

7. _____

8. $(2x^3 - 4x^2 + 3x + 8) \div \left(x - \dfrac{1}{2}\right)$

8. _____

Objective 2

For the given polynomial $P(x)$ and the given c, use the remainder theorem to find $P(c)$.

9. $P(x) = x^5 - 9x^4 + 3x^3 + x - 2;\ -2$

9. _____

10. $P(x) = 4x^4 + x^2 - 2;\ -1$

10. _____

11. $P(x) = 3x^5 + 4x^2 + 9x - 8$; 2

11. _____

12. $P(x) = 4x^4 + x^2 - 2$; -1

12. _____

13. $P(x) = 4x^4 + 2x^3 - 6x + 7$; -2

13. _____

14. $P(x) = 5x^4 - 4x^3 - 7x^2 - 5x + 1$; 4

14. _____

Concept Extension

Divide.

15. $\left(5x^4 + \dfrac{1}{3}x^3 + x - 1\right) \div (x + 2)$

15. _____

Chapter 5 Vocabulary

Exponent Rules:		
Power Rule:	$\left(x^m\right)^n = x^{m \cdot n}$	$\left(2^3\right)^4 = 2^{12}$
Product Rule:	$x^m \cdot x^n = x^{m+n}$	$2^3 \cdot 2^4 = 2^7$
Power of a Product Rule:	$\left(xy\right)^n = x^n y^n$	$\left(2y\right)^4 = 2^4 y^4$
Power of a Quotient Rule:	$\left(\dfrac{x}{y}\right)^n = \dfrac{x^n}{y^n}$	$\left(\dfrac{2}{y}\right)^4 = \dfrac{2^4}{y^4}$
Quotient Rule:	$x^m \div x^n = x^{m-n}$	$2^5 \div 2^2 = 2^3$
Zero Exponent:	$x^0 = 1$	$2^0 = 1$
Polynomial	Finite sum of terms in the form ax^2	$8x^3 + 3x^2 - 2x - 9$
Monomial	A one term polynomial	$6c$
Binomial	A two term polynomial	$6x - 4$
Trinomial	A three term polynomial	$3x^2 - 2x - 9$
Degree of term	Sum of the exponents on the variables	$3x^2 y^4$ $2 + 4 = 6$
Degree of polynomial	The greatest degree of terms of the polynomial	$3x^2 - 2x - 9$ Degree of terms were 2, 1, and 0. Therefore the degree of polynomial is 2.
FOIL	Method to multiply binomials. First, Outer Inner, Last	$(x-1)(x+2)$ F: x^2 O: +2x I: - x L = -2

Squaring a binomial	$(a \pm b)^2 = a^2 \pm 2ab + b^2$	$(x-1)^2 =$ $x^2 - 2(x) + (1)^2 =$ $x^2 - 2x + 1$
Multiplying a Sum and Difference	$(a+b)(a-b) = a^2 - b^2$	$(x-3)(x+3) =$ $x^2 - (3)^2 =$ $x^2 - 9$
Scientific Notation	Method to write extremely large numbers or small decimals in a simplified manner.	$1,250,000,000 = 1.25 \times 10^9$ $0.0000000657 = 6.57 \times 10^{-8}$
Synthetic Division	A shortcut method to divide a polynomial by a binomial of the form $x - c$	

Name: **Date:**

Instructor: **Section:**

Practice Test A

Evaluate each expression.

1. $\left(\dfrac{1}{8}\right)^3$

1. _____

2. -2^6

2. _____

Simplify each exponential expression. Write the results using only positive exponents.

3. $\dfrac{\left(r^{-2}s^3\right)^{-5}}{2r^{-5}s^4}$

3. _____

4. $\left(-3m^4n\right)^3\left(4m^3n^4\right)$

4. _____

Express each number in scientific notation.
5.   137,200,000,000

5. _____

Write each number in standard form.
6. 4.284×10^{-8}

6. _____

Simplify.

7. $\dfrac{1.5\times10^{-8}}{3\times10^{-9}}$

7. _____

Simplify by combining like terms.
8. $7x^2y+13xy^2-6x^2y+2y^2-8xy^2$

8. _____

Perform the indicated operation.
9. $\left(3a^2-4a-2\right)+\left(6a^2+10a\right)$

9. _____

Martin-Gay *Beginning and Intermediate Algebra, Fourth Edition*   167

10. Subtract $\left(7x^2 - 8x\right)$ from $\left(9 - 3x^2\right)$

10. _____

11. Subtract $\left(c^2 - 2c + 10\right)$ from the sum of $\left(3c^2 - 8c + 9\right)$ and $\left(-4c^2 + 8c - 5\right)$.

11. _____

Multiply.

12. $\left(a + b\right)\left(2a - 3b\right)$

12. _____

13. $-\dfrac{1}{4}x\left(16x^3 - 20x^2 + 8x - 32\right)$

13. _____

14. $\left(2m + n\right)\left(2m^2 - 8m + n\right)$

14. _____

15. $\left(x - 13\right)\left(x + 13\right)$

15. _____

16. $\left(y - 7\right)^2$

16. _____

17. $\left(7x - \dfrac{1}{2}\right)\left(7x + \dfrac{1}{2}\right)$

17. _____

18. $\left(c + 4\right)^3$

18. _____

19. Find the area of a rectangle if the width is five less than twice the length. Given that the length is x, state your answer in terms of x.

19. _____

20. The height of a falling object is given by the equation $-16x^2 + 324$, where x is time in seconds. How far does the object fall in 6 seconds?

20. _____

21. $\dfrac{4x^3 - 8x^2 - 6x}{2x}$

21. _____

22. $\dfrac{6m^2n - 8mn^2 - 16m}{8mn}$

22. _____

23. $\dfrac{2x^2 + x - 21}{x - 3}$

23. _____

24. $\dfrac{x^3 - 125}{x - 5}$

24. _____

25. If $P(x) = 4x^3 - 6x^2 + x - 12$, use the remainder theorem to find $P(-1)$.

25. _____

Practice Test B

Evaluate each expression.
1. -0.04^3

    a. 0.000064

    c. -0.000064

    b. -0.12

    d. 0.12

1. _____

2. $(-1)^6$

    a. 6

    c. -1

    b. -6

    d. 1

2. _____

Simplify each exponential expression. Write the results using only positive exponents.

3. $\dfrac{\left(x^5 y^{-4}\right)^{-3}}{5y^6 x^{-4}}$

    a. $\dfrac{x^{11} y^6}{5}$

    b. $\dfrac{x^{19}}{5y^{18}}$

    c. $\dfrac{x^{11}}{5y^{10}}$

    d. $\dfrac{y^6}{5x^{11}}$

3. _____

4. $\left(-7s^9 r\right)\left(-5s^2 r^3\right)^2$

    a. $35s^{11} r^4$

    c. $35s^{13} r^7$

    b. $70s^{13} r^6$

    d. $-175s^{13} r^7$

4. _____

Express each number in scientific notation.
5.   65,790,000,000

    a. 6.579×10^{-10}

    c. 6579×10^{-7}

    b. 6.579×10^{10}

    d. 6579×10^7

5. _____

Martin-Gay *Beginning and Intermediate Algebra, Fourth Edition*

Name:

Date:

Instructor:

Section:

Write each number in standard form.

6. 9.32×10^{-4}

    a. 93,200

    b. 9,320,000

    c. 0.000932

    d. 0.0000932

6. _____

Simplify.

7. $\left(1.2 \times 10^{12}\right)\left(6 \times 10^{-8}\right)$

    a. 7.2×10^{4}

    b. 5.0×10^{4}

    c. 7.2×10^{-4}

    d. 7.2×10^{20}

7. _____

Simplify by combining like terms.

8. $5x^2 - 3y^2 + 6x - 8y + 14x^2 - 16y$

    a. $19x^2 + 6x - 24y - 3y^2$

    b. $-9x^2 + 6x - 8y - 3y^2$

    c. $25x - 27y$

    d. $-4x^2 y^2$

8. _____

Perform the indicated operation.

9. $(7x - 4y) + (-13x - 8y)$

    a. $20x - 12y$

    b. $20x + 4y$

    c. $-6x - 12y$

    d. $-6x + 4y$

9. _____

10. Subtract $\left(12m - 8m^2\right)$ from $\left(11m^2 + 12m\right)$

    a. $3m^2$

    b. $3m^2 + 24m$

    c. $19m^2 + 24m$

    d. $19m^2$

10. _____

11. Subtract $\left(2x^2 - 5x + 13\right)$ from the sum of $\left(8x^2 - 11x - 7\right)$ and $\left(-13x^2 - 4x + 17\right)$.

    a. $9x^2 - 15x - 10$ b. $-7x^2 - 10x - 3$

11. _____

    c. $-5x^2 - 15x + 10$ d. $-7x^2 - 20x + 23$

Multiply.

12. $\left(x - 2y\right)\left(4x + 3y\right)$

    a. $4x^2 - 5xy - 6y^2$ b. $4x^2 + 5xy + 6y^2$

12. _____

    c. $4x^2 + 11xy - 6y^2$ d. $4x^2 - 11xy - 6y^2$

13. $-5x\left(2x^3 + 10x^2 - 11x + 16\right)$

    a. $-3x^4 + 5x^3 - 16x^2 - 11x$ b. $-10x^3 - 50x^2 + 55x - 90$

13. _____

    c. $-10x^4 - 50x^3 + 55x^2 - 90x$ d. $-3x^3 + 5x^2 - 16x - 11$

14. $\left(8x - 1\right)\left(2x^2 + 3x + 4\right)$

    a. $16x^3 + 26x^2 + 35x + 4$ b. $16x^3 + 26x^2 - 29x - 4$

14. _____

    c. $16x^3 - 22x^2 - 35x - 4$ d. $16x^3 + 22x^2 + 29x - 4$

15. $\left(2x - 7\right)\left(2x + 7\right)$

    a. $4x^2 - 49$ b. $4x^2 - 28x - 49$

15. _____

    c. $4x^2 + 49$ d. $4x^2 + 28x - 49$

16. $(w+3)^2$

     a. $w^2 + 9$ b. $w^2 + 6$

16. _____

     c. $w^2 + 6w + 9$ d. $w^2 + 6w + 6$

17. $(x - 0.25)(x + 0.25)$

     a. $x^2 + 0.50x + 0.0625$ b. $x^2 - 0.0625$

17. _____

     c. $x^2 + 0.0625$ d. $x^2 - 0.50x - 0.0625$

18. $(x - 2)^3$

     a. $x^3 - 6x^2 + 12x - 8$ b. $x^3 + 6x^2 + 12x + 8$

18. _____

     c. $x^3 - 8x - 8$ d. $x^3 - 16x^2 - 2x + 16$

19. Find the area of a triangle if the base is four less than double the height. Given that the height is x, state your answer in terms of x.

     a. $x^2 - 2$ b. $2x^2 - 4x$

     c. $2x^2 - 4$ d. $x^2 - 2x$

19. _____

20. The height of a falling object is given by the equation $-16x^2 + 1648$, where x is time in seconds. How far does the object fall in 8 seconds?

    a. -376 feet

    c. 1556 feet

    b. 520 feet

    d. 624 feet

20. _____

21. $\dfrac{16m^5 - 24m^2 - 8}{4m^2}$

    a. $4m^7 - 6m^4 - 2$

    b. $4m^3 - 6m - 2$

21. _____

    c. $4m^3 - 6 - \dfrac{2}{m^2}$

    d. $4m^3 + 6 + \dfrac{2}{m^2}$

22. $\dfrac{5x^4 - 7x^2 - 15x}{5x^2}$

    a. $x^2 - \dfrac{7}{5}x - 3$

    b. $x^2 + \dfrac{7}{5} + \dfrac{3}{x}$

22. _____

    c. $x^2 - \dfrac{7}{5} - \dfrac{3}{x}$

    d. $x^4 - \dfrac{7}{5}x^2 - 3x$

23. $\dfrac{8x^3 - 1}{2x - 1}$

    a. $4x^2 + 1$

    b. $4x^2 - 2x - 1$

                       23. _____

    c. $4x^2 - 2x + 1$

    d. $4x^2 + 2x + 1$

24. $\left(4x^2 - 29x + 40\right) \div (x - 6)$

    a. $4x - 5 + \dfrac{10}{x - 6}$

    b. $4x - 5$

                       24. _____

    c. $4x + \dfrac{-5x + 40}{x - 6}$

    d. $4x + 5 - \dfrac{10}{x - 6}$

25. If $P(x) = 6x^3 + 4x^2 - 5x + 11$, use the remainder theorem to find $P(-2)$.

    a. -31

    b. -11

                       25. _____

    c. 85

    d. 65

Chapter 6 Factoring Polynomials
Section 6.1 The Greatest Common Factor and Factoring by Grouping

---

Learning Objectives
1. Find the greatest common factor of a list of integers.
2. Find the greatest common factor of a list of terms.
3. Factor out the greatest common factor from a polynomial.
4. Factor a polynomial by grouping.

Vocabulary
Use the choices to complete each statement.

Factoring	**Factors**	**False**
Greatest common factor	**True**	

1. _____ is what is multiplied together to result in a product.

2. True or False. A number that does not have any factors other than one and itself is a composite number. _____.

3. The _____ is the largest factor that can divide into all the terms evenly.

4. _____ is the mathematical process of rewriting a polynomial as a product.

Objective 1

Find the GCF for each list.

5. 18, 27, 36

5. _____

6. 64, 82, 108

6. _____

7. 45, 39, 19

7. _____

Objective 2

Find the GCF for each list

8. m^8, m^6, m^4

8. _____

9. $12y^4$ and $20y^3$

9. _____

10. $-40x^2y^3$, $20x^4y^2$, $16x^3y^3$

10. _____

11. $15a^7b^5$, $25a^4b^3$, $27a^8b^9$

11. _____

Objective 3

Factor out the GCF from each polynomial.

12. $14x - 7$

12. _____

13. $9x^2 - 18x + 6$

13. _____

14. $m^4n^3 - m^3n^3 + m^5n^4 - m^6n^7$

14. _____

15. $y(x^2 + 2) + 3(x^2 + 2)$

15. _____

Objective 4

Factor the given polynomials by grouping.

16. $5xy - 15x - 6y + 18$

16. _____

17. $ab + 5b^2 - 4a - 20b$

17. _____

18. $21xy - 56x + 6y - 16$

18. _____

Concept Extension

Factor.

19. $168ab + 96a - 189b - 108$

19. _____

20. $24y^{2n} - 60y^n - 20y^n + 50$

20. _____

Section 6.2 Factoring Trinomials of the Form $x^2 + bx + c$

Learning Objectives
1. Factor trinomial of the form $x^2 + bx + c$
2. Factor out the greatest common factor and then factor trinomial of the form $x^2 + bx + c$

Objective 1

Factor each trinomial.

1. $x^2 - 3x - 18$

1. _____

2. $x^2 - 11x + 28$

2. _____

3. $x^2 + 12x + 32$

3. _____

4. $x^2 - 13x - 36$

4. _____

5. $x^2 - 2x - 3$

5. _____

6. $x^2 - 5x + 6$

6. _____

7. $x^2 - xy - 20y^2$

7. _____

Objective 2

Factor each trinomial completely.

8. $3x^2 + 9x - 30$

8. _____

9. $x^2 - 7x + 44$

9. _____

10. $5x^3y - 25x^2y^2 - 120xy^3$

10. _____

11. $16x^2y - 16xy - 180y$

11. _____

12. $-4x^4 + 8x^2 + 60$

12. _____

13. $\frac{1}{2}x^2 - 2x - 16$

13. _____

Concept Extension

Factor completely.

14. $t^2(s-1) - 3t(s-1) - 28(s-1)$

14. _____

15. Find a positive value for b so that this trinomial is factorable. $x^2 + bx - 44$

15. _____

16. Find a positive value for c so that this trinomial is factorable. $x^2 - 17x + c$

16. _____

Section 6.3 Factoring Trinomials of the Form $ax^2 + bx + c$ and Perfect Square Trinomials

Learning Objectives
1. Factor trinomial of the form $ax^2 + bx + c$ where $a \neq 1$
2. Factor out the greatest common factor then factor trinomial of the form $ax^2 + bx + c$
3. Factor a perfect square trinomial

Objective 1

Factor completely.

1. $6x^2 + x - 2$

1. _____

2. $10x^2 + 31x + 3$

2. _____

3. $5x^2 - 14x + 20$

3. _____

4. $3x^2 - 19x - 14$

4. _____

5. $27x^2 - 57x + 20$

5. _____

Objective 2

Factor completely.

6. $6x^2 - 27x - 15$

6. _____

7. $4x^3 - 9x^2 - 9x$

7. _____

8. $-16x^2 + 80x - 100$

8. _____

9. $56m^2 - 63 + 7m$

9. _____

10. $36x^2y + 21xy - 30y$

10. _____

Objective 3

Factor completely.

11. $16x^2 - 40x + 25$

11. _____

12. $9x^2 - 24xy + 16y^2$

12. _____

13. $16x^2 - 48x + 36$

13. _____

14. $64x^2 - 144x + 81$

14. _____

Concept Extension

Factor completely.

15. $36x^2 - 6x + \dfrac{1}{4}$

15. _____

16. $\dfrac{16}{49}x^2 - \dfrac{8}{21}xy + \dfrac{1}{9}y^2$

16. _____

17. $\dfrac{1}{4}x^2 - \dfrac{2}{3}x + \dfrac{1}{3}$

17. _____

Section 6.4 Factoring Trinomials of the Form $ax^2 + bx + c$ by Grouping

Learning Objective
1. Use the grouping method to factor trinomial of the form $ax^2 + bx + c$

Objective 1

Factor completely.

1. $x^2 + 3x + 2x + 6$

1. _____

2. $30x^2 - 18x - 5x + 3$

2. _____

3. $2a^2 - 5a - 8a + 20$

3. _____

4. $4x^2 - 9x - 9$

4. _____

5. $12x^2 - 8x - 7$

5. _____

6. $9x^2 - 25x + 20$

6. _____

7. $36x^2 - 60x - 24$

7. _____

8. $20x^2 y + 22xy - 12y$

8. _____

9. $12x^2 + 47xy + 40y^2$

9. _____

10. $21x^2 - 17x - 8$

10. _____

Concept Extension

Factor completely.

11. $\dfrac{1}{2}x^2 + \dfrac{11}{4}x - \dfrac{3}{2}$

11. _____

12. $x^2(x-7) - 16(x-7)$

12. _____

Section 6.5 Factoring Binomials

Learning Objectives
  1. Factor the difference of two squares.
  2. Factor the sum or difference of two cubes.

Vocabulary.
Use the choices to complete each statement.
Difference of cubes   **Difference of squares**
Perfect square trinomial   **Sum of cubes**

1. $25x^2 - 36$ is an example of _____.

2. $8x^3 + 64$ is an example of _____.

3. $(x+1)^2$ when FOILed out will result in a _____.

4. $125x^3 - 1$ is an example of _____.

Objective 1

Factor completely.

5. $x^2 - 49$

5. _____

6. $25x^2 - 144$

6. _____

7. $121m^2 - 100n^2$

7. _____

8. $81x^2 - 64$

8. _____

9. $169x^2 + 36y^2$

9. _____

10. $x^8 - 1$

10. _____

11. $\dfrac{25}{36}t^2 - \dfrac{4}{9}$

11. _____

Objective 2

Factor completely.

 12. $x^3 + 125$

12. _____

13. $y^3 + 27$

13. _____

14. $b^3 - 216$

14. _____

15. $t^3 + 16$

15. _____

 16. $xy^3 - 9xyz^2$

16. _____

17. $x^4 - y^6$

17. _____

18. $64t^3 - 729s^3$

18. _____

Concept Extension

Factor completely.

19. $(x-1)^3 - 27$

19. _____

20. $2.25t^2 - 1.69$

20. _____

Section 6.6 Solving Quadratic Equations by Factoring

Learning Objectives
1. Solve quadratic equation by factoring.
2. Solve equations with degree greater than two by factoring.
3. Find the x-intercept of a quadratic equation in two variables.

Objective 1

Solve the following equations.

1. $x^2 + x - 12 = 0$

1. _____

2. $2x^2 + 7x - 15 = 0$

2. _____

3. $6x^2 - 17x = -7$

3. _____

4. $2x^2 - 6x - 20 = 0$

4. _____

5. $x(x + 14) = -49$

5. _____

6. $6x^2 + x - 15 = 0$

6. _____

Objective 3

Factor completely.

7. $5x^3 - 5x = 0$

7. _____

8. $(2x+3)(4x-5) = 0$

8. _____

9. $x^3 - 5x^2 - 24x = 0$

9. _____

10. $y^3 + y^2 - 6y = 0$

10. _____

11. $-3x^3 + 6x^2 + 9x = 0$

11. _____

Objective 3

Find the x-intercept of the following quadratic equations.

12. $y = x^2 - 4x - 21$

12. _____

13. $y = 2x^2 + 11x - 6$

13. _____

14. $y = 6x^2 - 5x - 4$

14. _____

15. $y = 3x^3 - 24x^2$

15. _____

Concept Extension

Solve.

16. An object is dropped from a height of 560 feet. The height of the object is given by the equation $h(t) = -16t^2 - 32t + 560$, at what time, t, will the object hit the ground?

16. _____

Section 6.7 Quadratic Equations and Problem Solving

Learning Objectives
  1 . Solve problems that can be modeled by quadratic equations.

Objective 1

Solve.

1. The perimeter of the triangle is 85 feet. Find the lengths of its sides.

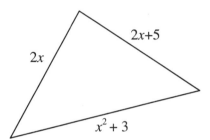

          1. _____

2. The area of a square is 169 square units. Find the length of its sides.

          2. _____

3. The product of two consecutive even integers is 168. Find the numbers.

          3. _____

4. The sum of a number and its square is 182. Find the number(s).

          4. _____

5. The difference of two integers is 2, and their product is 143. Find the numbers.

5. _____.

6. An object is thrown upward from the top of an 80-foot building with an initial velocity of 64 feet per second. The height h of the object after t seconds is given by the quadratic equation $h = -16t^2 + 64t + 80$. When will the object hit the ground?

6. _____

7. The perimeter of a textbook is 28 inches. And the diagonal measures 10 inches. What are the length and width of the textbook?

7. _____

8. Melissa's age is twice Alice's. If the sum of the squares of their ages is 80, then what are their ages?

8. _____

9. The diameter of a circle is two less than the diameter of another circle. If the areas of the circles differ by 5π square meters, then what is the diameter of each circle?

9. _____

10. The length of the longer side of a right triangle is 2 feet less than the hypotenuse. The length on the shorter leg is 7 feet less than the length of the longer leg. Find the length of all three sides.

10. _____

Concept Extension

11. The sum of two numbers is 28. The sum of their squares is 400. Find the two numbers.

11. _____

12. Write down two numbers whose sum is 25. Square each number, and find the sum of their squares. Use this work to create your own word problem, like in number 11.

12. _____

Chapter 6 Vocabulary

Vocabulary Word	Definition	Example
Factoring	The process of writing an expression as a product.	$10 = 5 \cdot 2$ $x^2 + 3x + 2 = (x+2)(x+1)$
Greatest Common Factor	The biggest factor that goes into all the other terms.	2 is the GCF of 4, 18, and 22 x^2 is the GCF of x^2, x^4, x^7
Factor by grouping	Method to factor a polynomial, find the GCF between pairs of terms, then factor out a common binomial if present.	$2x^3 + x^2 + 6x + 3$ $x^2(2x+1) + 3(2x+1)$ $(2x+1)(x^3+3)$
Perfect Square Trinomial	$a^2 \pm 2ab + b^2$	$4x^2 - 12x + 9$
Difference of Squares	$a^2 - b^2$	$16x^2 - 25$
Sum of Cubes	$a^3 + b^3$	$8x^3 + 125$
Difference of Cubes	$a^3 - b^3$	$64x^3 - 1$
Quadratic Equation	$ax^2 + bx + c = 0$	$2x^2 - 3x + 2 = 0$
Zero Factor Theorem	If $a \cdot b = 0$, then $a = 0$ or $b = 0$.	$(x-2)(x+1) = 0$ $x - 2 = 0 \qquad x + 1 = 0$ $x = 2 \qquad\quad x = -1$

Practice Test A

Factor each polynomial completely.

1. $x^2 - 3x - 28$

1. _____

2. $6x^2 + 11x - 10$

2. _____

3. $8x^3 + 1$

3. _____

4. $3x^2 - 30x + 27$

4. _____

5. $x^4 - 16$

5. _____

6. $-4x^2 + 23x - 28$

6. _____

7. $3x^3 - 6x^2 - 5x + 10$

7. _____

8. $4x^2 - 169$

8. _____

9. $x^2 + 11x + 30$

9. _____

10. $16a^2 + 25$

10. _____

Solve.

11. $x^2 - 13x + 36 = 0$

11. _____

12. $2x^3 + 14x^2 + 12x = 0$

12. _____

13. $20x^2 + 26x - 6 = 0$

13. _____

14. $-15x^2 - 16x = 4$

14. _____

15. $25x^2 = -16 - 40x$

15. _____

16. $24x^2 - 20x - 4 = 0$

16. _____

Find the x-intercept of the given quadratic equation.

17. $y = 11x^2 + 21x - 2$

17. _____

18. $y = 100x^2 - 40x + 4$

18. _____

19. The difference of two numbers is 5. The sum of their squares is 125. Find the numbers.

19. _____

20. A compass is falling off a 324 foot cliff. If the height of the compass can be found using the equation, $h = -16t^2 + 324$, where t is time in seconds. At what time will the compass hit the ground?

20. _____

Practice Test B

Factor each polynomial completely.

1. $x^2 + 4x - 21$

    a. $(x+7)(x-3)$ b. $(x-7)(x+3)$

                                                    1. _____

    c. $(x - 7)(x - 3)$ d. prime

2. $2x^2 + 9x - 18$

    a. $(2x - 9)(x + 2)$ b. $(x + 6)(2x - 3)$

                                                    2. _____

    c. $(x + 9)(2x-2)$ d. $(x - 6)(2x + 3)$

3. $125 - 64x^3$

    a. $(5 - 4x)(5+4x)^2$ b. $(5-4x)(25 - 20x + 16x^2)$

                                                    3. _____

    c. $(5-4x)(25 + 20x + 16x^2)$ d. prime

4. $x^3 + 3x^2 - 10x$

    a. $x(x - 2)(x + 5)$ b. $(x^2 - 2x)(x+5)$

                                                    4. _____

    c. $(x - 2)(x^2+5x)$ d. prime

5. $y^2 - 11y + 30$

    a. $(y - 3)(y - 10)$ b. $(y - 5)(y - 6)$

                                                    5. _____

    c. $(y + 5)(y + 6)$ d. prime

6.. $3x^2 - 12$

   a. $3(x^2-4)$ b. $3(x-2)^2$

6. _____

   c.prime   d. $3(x+2)(x-2)$

7. $12x^2 + 54x - 30$

   a. $(4x-3)(3x-10)$ b. $6(x+5)(2x-1)$

7. _____

   c. $6(2x-5)(x-1)$ d. prime

8. $x^2 y + 3xy - 4y$

   a. $y(x-1)(x+4)$ b. $y(x-2)(x+2)$

8. _____

   c. $(xy-2)(x-2y)$ d. prime

9. $4x^2 - 4xy - 24y^2$

   a. $4(x-3y)(x+2y)$ b. $2(x-2y)(x+6y)$

9. _____

   c. $4(x-4y)(x-3y)$ d. prime

10. $15 - 5x - 3x^3 + 9x^2$

   a. $(x-3)(5-3x^2)$ b. $(x^2-3)(5-x)$

10. _____

   c. $3(x+5)(x-2)$ d. prime

Solve.

11. $2x^3 - 2x^2 - 40x = 0$

    a.   0, -10, 4   b. -5, -1, and 8

                                                       11. _____

    c. 0, 5, -4 d. no solution

12. $4x^2 - 25 = 0$

    a. 5 and -5   b. 2 and - 2

                                                     12. _____

    c. $-\dfrac{5}{2}$ and $\dfrac{5}{2}$ d. no solution

13. $3x^2 + 8x + 5 = 0$

    a. 1, and 5   b. $-\dfrac{5}{3}$ and -1

                                                      13. _____

    c. -1 and -3   d. no solution

14. $3x^3 - 2x^2 - 8x = 0$

    a. 1, -3 and 8   b.1, 2, and 4

                                                      14. _____

    c. no solution   d. 0, 2, and $-\dfrac{4}{3}$

15. $18y^3 + 3y^2 - 3y = 0$

    a. 6, -3, and 1

    b. 0, $-\frac{1}{2}$, and $\frac{1}{3}$

15. _____

    c. 1, $-\frac{1}{4}$, and $\frac{2}{3}$

    d. no solution

16. $x^2 + 7x + 12 = 0$

    a. -12 and -1

    b. - 4, - 3

16. _____

    c. 6 and 2

    d. no solution

Find the *x*-intercept of the given quadratic equation.

17. $y = 10x^2 + 41x + 21$

    a. -3 and -7

    b. 7 and 3

17. _____

    c. $-\frac{3}{5}$ and $-\frac{7}{2}$

    d. no x-intercept

18. $y = 6x^2 - 2x - 28$

    a. 2 and 7

    b. -7 and 4

18. _____

    c. no x-intercept

    d. $\frac{7}{3}$ and - 2

19. The area of a square is 144 square units. Find the length of a side

    a. 13   b. 12

    c. 14   d. 10

19. _____

20. Find two consecutive negative integers whose product is 240. Find the numbers

    a. -12 and - 20   b. – 24 and - 10

    c. -15 and -16   d. 12 and 20

20. _____

Chapter 7 Rational Expressions
Section 7.1 Rational Functions and Simplifying Rational Expressions

Learning Objectives
1.　　Find the domain of a rational expression.
2.　　Simplify rational expressions.
3.　　Write equivalent forms of rational expressions.
4.　　Use rational functions in applications.

Vocabulary
Use the choices to complete each statement.

Domain	**Rational**	**Simplified**
1	**-1**	**0**

1. The expression $\dfrac{x-y}{y-x}$ will simplify down to _____.

2. Like a fraction, if the denominator of a rational expression is _____ then the expression in undefined.

3. The _____ of the function $f(x) = \dfrac{6}{x+4}$ is $\{x \mid x \text{ is a real number}, x \neq -4\}$.

4. A _____ expression is an expression that can be written as a quotient of two polynomials, but the denominator cannot be 0.

5. The expression $\dfrac{a+b}{b+a}$ will simplify down to _____.

Objective 1
Find the domain of each rational expression.

6. $f(x) = \dfrac{3x+1}{x-5}$　　　　　　　　　　　　　　6. _____

7. $g(x) = \dfrac{4x}{x^2 + 3x + 2}$　　　　　　　　　　　7. _____

8. $h(x) = \dfrac{5x}{3x+2}$　　　　　　　　　　　　　　8. _____

9. $C(x) = \dfrac{x+3}{x^2-4}$

9. _____

Objective 2

Simplify each rational expression.

10. $\dfrac{6y-18}{8y-24}$

10. _____

11. $\dfrac{x^3+7x^2}{x^2+5x-14}$

11. _____

12. $\dfrac{-a+b}{b-a}$

12. _____

13. $\dfrac{x^2-9}{x^2+x-6}$

13. _____

Objective 3

List four equivalent forms for each rational expression.

14. $-\dfrac{x+11}{x-4}$

14. _____

15. $-\dfrac{2y-5}{7-3y}$

15. _____

16. $-\dfrac{x-5}{9-x}$

16. _____

Objective 4

17. The total revenue from the sale of a popular book is approximately nu the rational expression
$R(x) = \dfrac{1000x^2}{100-x}$, where x is the number of years since publication and $R(x)$ is the total revenue in millions of dollars.

    a.   Find the total revenue at the end of the first year.

17a. _____

    b.   Find the total revenue at the end of the second year.

17b. _____

    c.   Find the revenue during the second year only.

17c. _____

    d.   Find the domain of function R.

17d. _____

Concept Extension

18. Is the expressions $x-7$ and $\dfrac{x^2-9x-14}{x-2}$ equivalent for all real numbers? If not what values, would the two expressions not be equivalent.

18. _____

19.   Find the area of a triangle with height of $\dfrac{x}{x+3}$ and base of $\dfrac{x^2+6x+9}{2x}$.

19. _____

Section 7.2 Multiplying and Dividing Rational Expressions

Learning Objectives
1. Multiply rational expressions.
2. Divide rational expressions.
3. Multiply or divide rational expressions.

Objective 1

Find each product and simplify if possible.

1. $\dfrac{2w^2}{15x^3} \cdot \dfrac{25x}{6w}$

1. _____

2. $\dfrac{16t^2}{35} \cdot 70t^4$

2. _____

3. $\dfrac{x-2}{x+3} \cdot \dfrac{4}{x-2}$

3. _____

4. $\dfrac{4x^2-1}{2x^2+x-1} \cdot \dfrac{5x^2-5}{16x+8}$

4. _____

5. $\dfrac{z^2-1}{(z-1)^2} \cdot \dfrac{z-1}{z^2+2z+1}$

5. _____

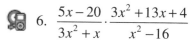 6. $\dfrac{5x-20}{3x^2+x} \cdot \dfrac{3x^2+13x+4}{x^2-16}$

6. _____

Martin-Gay *Beginning and Intermediate Algebra, Fourth Edition*

Objective 2

Find each quotient and simplify.

7. $\dfrac{49x^2}{16y^3} \div \dfrac{28x^4}{32y^5}$

7. _____

8. $\dfrac{(x+3)(x-3)}{(x-7)(x-2)} \div \dfrac{x-7}{x-2}$

8. _____

9.

$$\dfrac{\dfrac{x+2}{x-3}}{\dfrac{x+2}{x-4}}$$

9. _____

10. $\dfrac{x+2}{7-x} \div \dfrac{x^2-5x+6}{x^2-9x+14}$

10. _____

11. Find the quotient of $\dfrac{x^2-1}{x-4}$ and $\dfrac{x^2+2x+1}{x^2-7x+12}$.

11. _____

Objective 3

Multiply or divide as indicated.

12. $\dfrac{4x^2-9}{x^2+x-12} \cdot \dfrac{x^2+-x-6}{4x^2+12x+9}$

12. _____

13. $\dfrac{6x^2+5x-4}{2x^2+9x-5} \div \dfrac{3x^2-8x-16}{x^2+x-20}$

13. _____

14. Find the quotient of $\dfrac{x^2-7x-30}{x-5}$ and $\dfrac{x^2-6x-40}{x^2-10x+25}$.

14. _____

Concept Extension

Multiply or divide as indicated.

15. $\left(\dfrac{x^2+4x+3}{x^2+7x+10} \cdot \dfrac{x^2+4x-5}{x^2+2x-3} \right) \div \left(\dfrac{x^2+5x+4}{x^2+5x+6} \div \dfrac{x^2+2x-8}{x^2+2x-3} \right)$

15. _____

Section 7.3 Adding and Subtracting Rational Expressions with Common Denominators and Least Common Denominator

Learning Objectives
1. Add and subtract rational expressions with the same denominator.
2. Find the least common denominator of a list of rational expressions.
3. Write a rational expression as an equivalent expression whose denominator is given.

Objective 1

Add or subtract as indicated. Simplify if possible.

1. $\dfrac{x-3}{x+2}+\dfrac{2x-7}{x+2}$

1. _____

2. $\dfrac{-10}{x-6}+\dfrac{x+4}{x-6}$

2. _____

3. $\dfrac{x-4}{x+3}-\dfrac{7x}{x+3}$

3. _____

4. $\dfrac{2x+3}{x^2-x-30}-\dfrac{x-2}{x^2-x-30}$

4. _____

5. $\dfrac{x^2-2x+3}{x+1}-\dfrac{x^2+3x-5}{x+1}$

5. _____

6. $\dfrac{2x^2+4x+3}{5x+15}+\dfrac{3x^2+x+2}{5x+15}$

6. _____

Objective 2

Find the LCD for each list of rational expressions.

7. $\dfrac{4}{x-3}, \dfrac{5}{x+1}$

7. _____

8. $\dfrac{x-2}{x^2+2x+1}, \dfrac{x+3}{x^2+5x+4}$

8. _____

9. $\dfrac{5x}{x^2-16}, \dfrac{7x+1}{x^2+8x+16}$

9. _____

 10. $\dfrac{1}{3x+3}, \dfrac{8}{2x^2+4x+2}$

10. _____

11. $\dfrac{3x+1}{x^3+1}, \dfrac{2x-7}{x^2-1}$

11. _____

Objective 3

Rewrite each rational expression as an equivalent rational expression with the given denominator.

12. $\dfrac{7}{a+1} = \dfrac{}{a^2-1}$

12. _____

13. $\dfrac{3x}{x(x-2)} = \dfrac{}{x^3 - 5x^2 + 6x}$

13. _____

14. $\dfrac{9a+2}{5a+10} = \dfrac{}{5b(a+2)}$

14. _____

15. $\dfrac{x-1}{x^2+3x-10} = \dfrac{}{x^3+5x^2-4x-20}$

15. _____

Concept Extension

16. Write two rational expressions whose differences is $\dfrac{3x-9}{x^2-8}$.

16. _____

Section 7.4 Adding and Subtracting Rational Expressions with Unlike Denominators

| **Learning Objectives** |
| 1. Add and subtract rational expressions with unlike denominators. |

Objective 1

Add or subtract as indicated.

1. $\dfrac{4}{x+1}+\dfrac{3x}{x^2-1}$

1. _____

2. $\dfrac{8a}{a-2}+\dfrac{a+1}{2a-4}$

2. _____

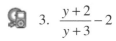 3. $\dfrac{y+2}{y+3}-2$

3. _____

4. $\dfrac{3x+1}{x-1}-\dfrac{2x-3}{1-x}$

4. _____

5. $\dfrac{x-3}{x^2+6x+9}+\dfrac{x+9}{x^2+8x+15}$

5. _____

6. $\dfrac{x+1}{x^2+2x-48} - \dfrac{2x}{x^2-14x+48}$

6. _____

7. $\dfrac{2}{y^2+3y-4} - \dfrac{4}{y^2-y-12}$

7. _____

8. $\dfrac{4}{2x^2-7x-15} + \dfrac{x-2}{2x^2+17x+21}$

8. _____

9. $\dfrac{x+2}{x^2-7x-8} + \dfrac{x+1}{x^2-9x+8}$

9. _____

10. $\dfrac{y-8}{y^2-8y+12} - \dfrac{y+2}{y^2+4y-12}$

10. _____

11. $\dfrac{x+8}{x^2-5x-6} + \dfrac{x+1}{x^2-4x-5}$

11. _____

12. $\dfrac{x-3}{x^2-5x-14} - \dfrac{x+2}{x^2-10x-21}$

12. _____

Concept Extension

13. $\dfrac{3x+1}{x^2-8x+12} - \dfrac{2x}{x^2-4x-12} + \dfrac{4}{x^2-4}$

13. _____

Section 7.5 Solving Equations Containing Rational Expressions

Learning Objectives
1. Solve equations containing rational expressions.
2. Solve equations containing rational expressions for a specified variable.

Objective 1

Solve each equation and check each proposed solution.

1. $\dfrac{2}{x-1} = 5$

1. _____

2. $\dfrac{7}{x-2} + 2 = \dfrac{5}{x-2}$

2. _____

3. $\dfrac{t}{t-4} = \dfrac{t+4}{6}$

3. _____

4. $-1 + \dfrac{2y}{y+3} = \dfrac{-4}{y+4}$

4. _____

5. $\dfrac{x}{x-6} = \dfrac{6}{x-6} + 3$

5. _____

6. $\dfrac{2}{2n-3} - \dfrac{2}{10n^2 - 13n - 3} = \dfrac{n}{5n+1}$

6. _____

7. $\dfrac{2}{4x-3} + \dfrac{1}{3x+5} = \dfrac{7x+2}{12x^2 + 11x - 15}$

7. _____

8. $\dfrac{2x}{6x^2 + 7x - 3} + \dfrac{5}{2x^2 + 11x + 12} = \dfrac{x-3}{3x^2 + 11x - 4}$

8. _____

9. $\dfrac{y-2}{9y^2 - 1} - \dfrac{7}{6y-2} = -\dfrac{1}{3y-1}$

9. _____

10. $\dfrac{4r-4}{r^2 + 5r - 14} + \dfrac{2}{r+7} = \dfrac{1}{r-2}$

10. _____

Objective 2

Solve each equation for specified variable.

11. $\dfrac{1}{p} + \dfrac{1}{q} = \dfrac{1}{r}$ for p.

11. _____

12. $\dfrac{P - 2L}{W} = 2$ for L.

12. _____

13. $T = \dfrac{2U}{B + E}$ for B.

13. _____

14. $L = \dfrac{S - 2WH}{2(W + H)}$ for H.

14. _____

Concept Extension

15. Why do you have to check your proposed solutions? Why are they not solutions of the equations?

15. _____

Section 7.6 Proportion and Problem Solving with Rational Equations

Learning Objectives
1. Solve proportions
2. Use proportions to solve problems
3. Solve problems about numbers
4. Solve problems about work.
5. Solve problems about distance.

Objective 1

Solve each proportion.

1. $\dfrac{x}{6} = \dfrac{11}{2}$

1. _____

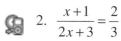

2. $\dfrac{x+1}{2x+3} = \dfrac{2}{3}$

2. _____

3. $\dfrac{3}{x+1} = \dfrac{4}{x-1}$

3. _____

4. $\dfrac{8}{x-4} = \dfrac{2}{x+2}$

4. _____

Objective 2

Solve.

5. In a class of 50 students, 30 of them are women. In a class, of 100 students, how many are expected to be women?

5. _____

6. It has been advertised that 3 out of 5 dentists endorse the Great Whiting toothpaste. If there are 1525 dentists at this year's annual New Hampshire dentist convention, how many of them endorse the toothpaste?

6. _____

7. If there is 1000 meters in 1 kilometer, then how many meters are in 4.56 kilometers?

7. _____

Objective 3

Solve.

8. The quotient of 5 and the sum of a number and 3 is the same as 8 divided by 7 more than that number. Find the number.

8. _____

9. Twelve divided by the sum of x and two equals the quotient of 4 and the difference of x and 2. Find x.

9. _____

10. The ratio of 4 and the first of two consecutive even integers equals the ratio of 6 and the second integer. Find the two integers.

10. _____

Objective 4

Solve.

11. In 2 minutes, a conveyor belt moves 300 pounds of recyclable aluminum from the delivery truck to a storage area. A smaller belt moves the same quantity of cans the same distance in 6 minutes. If both belts are used, find how long it takes to move the cans to the storage area.

11. _____

12. Paul paints a room in 6 hours. Joe can paint a room in 3 hours. How long will it take them to paint a room together?

12. _____

13. It takes Joel 2.5 hours to mow the yard, when his wife helps they complete the task in 1.5 hours. How long will it take Joel's wife to mow the yard by herself.

13. _____

Objective 5

Solve.

14. A car travels 280 miles in the same time a motorcycle travels 240 miles. If the car's speed is 10 miles per hour more than the motorcycle's, find the speed of the car and the speed of the motorcycle.

14. _____

15. Michelle can walk 16 miles in the same time it takes Lucy to walk 8 miles. If Michelle walks 1 mph faster than Lucy, how fast does each person walk?

15. _____

16. A Cessna-172 will leave Chicago and fly 450 miles to it's destination. A car will leave Chicago at the same time going to the same destination. If the Cessna-172 maintains a speed of 3 times that of the car, the Cessna-172 will reach the destination 6 hours ahead of the car. Find the speed of the Cessna-172.

16. _____

Section 7.7 Simplifying Complex Fractions

Learning Objective

1.   Simplify complex fractions by simplifying the numerator and denominator and then dividing.
2.   Simplify complex fractions by multiplying by a common denominator.
3.   Simplify expressions with negative exponents.

Objective 1

Simplify each complex fraction.

1. $\dfrac{\dfrac{2}{5x}}{\dfrac{4}{15x}}$ 1. _____

2. $\dfrac{\dfrac{3}{x+2}}{\dfrac{5}{x-2}}$ 2. _____

3. $\dfrac{\dfrac{4x^2-y^2}{xy}}{\dfrac{2}{y}-\dfrac{1}{x}}$ 3. _____

4. $\dfrac{\dfrac{4}{x} - \dfrac{3}{x^2}}{\dfrac{6}{x^3} + \dfrac{4}{x^2}}$

4. _____

Objective 2
Simplify each complex fraction.

5. $\dfrac{3 - \dfrac{4}{y}}{2 + \dfrac{2}{y}}$

5. _____

6. $\dfrac{\dfrac{x+2}{x} - \dfrac{2}{x-1}}{\dfrac{x+1}{x} + \dfrac{x+1}{x-1}}$

6. _____

7. $\dfrac{\dfrac{2}{xy} - 4}{\dfrac{5}{x} + \dfrac{3}{y}}$

7. _____

Martin-Gay *Beginning and Intermediate Algebra, Fourth Edition*

8. $\dfrac{\dfrac{-5x}{x+y}}{\dfrac{x}{y^2}}$

8. _____

Objective 3

Simplify.

9. $\dfrac{2a^{-1}+3b^{-2}}{a^{-1}-b^{-1}}$

9. _____

10. $\dfrac{4+y^{-1}}{3y^{-1}-2y^{-2}}$

10. _____

11. $\dfrac{5-(3x)^{-2}}{3+x^{-3}}$

11. _____

Concept Extension

Simplify.

12. $\dfrac{(x-1)^{-2} + 3x^{-1}}{5x^{-2} + (x-1)^{-2}}$

12. _____

Chapter 7 Vocabulary

Vocabulary Word	Definition	Example
Rational Expression	An expression that can be written as $\dfrac{P}{Q}$ where both P and Q are polynomials.	$\dfrac{x-1}{x+3}$
Multiply Rational Expressions	1. Factor numerators and denominators. 2. Multiply across 3. Simplify if needed.	$\dfrac{x-2}{x+3} \cdot \dfrac{2x}{x-1} = \dfrac{2x(x-2)}{(x+3)(x-1)}$
Divide Rational Expressions	Multiply by the reciprocal of the second rational expression.	$\dfrac{x-2}{x+3} \div \dfrac{2x}{x-1} = \dfrac{x-2}{x+3} \div \dfrac{x-1}{2x} =$ $\dfrac{2x(x-2)}{(x+3)(x-1)}$
Add or Subtract Rational Expressions	1. Factor the denominators. 2. Find an LCD. 3. Rewrite as equivalent fraction with the LCD. 4. Add or subtract. 5. Simplify if needed	$\dfrac{x-2}{x+3} + \dfrac{2x}{x-1} \qquad LCD = (x+3)(x-1)$ $\dfrac{(x-2)(x-1)}{(x+3)(x-1)} + \dfrac{2x(x+3)}{(x+3)(x-1)} =$ $\dfrac{x^2-3x+2+2x^2+6x}{(x+3)(x-1)} = \dfrac{3x^2+3x+2}{(x+3)(x-1)}$
Ratio	Quotient of two numbers	$\dfrac{3}{5}$ = 3 to 5
Proportion	A Mathematical statement that two ratios are equal.	$\dfrac{1}{4} = \dfrac{x}{16}$
Simplify Complex Fraction -Method I	1. Add or subtract the numerator and the denominator. 2. Perform the division 3. Simplify if needed.	$\dfrac{\dfrac{1}{x}+\dfrac{1}{2}}{\dfrac{1}{x}-\dfrac{1}{4}} = \dfrac{\dfrac{2+x}{2x}}{\dfrac{4-x}{4x}} =$ $\dfrac{2+x}{2x} \cdot \dfrac{4x}{4-x} = \dfrac{4x(2+x)}{2x(4-x)} = \dfrac{2(2+x)}{4-x}$

Simplify Complex Fraction -Method II	1. Find LCD of all fractions. 2. Multiply numerator and denominator by that LCD. 3. Add/Subtract 4. Simplify if needed	$\dfrac{\dfrac{1}{x}+\dfrac{1}{2}}{\dfrac{1}{x}-\dfrac{1}{4}}$ $LCD=4x$ $\dfrac{4x\left(\dfrac{1}{x}+\dfrac{1}{2}\right)}{4x\left(\dfrac{1}{x}-\dfrac{1}{4}\right)}=\dfrac{4+2x}{4-x}=\dfrac{2(2+x)}{4-x}$

Practice Test A

Find any real numbers for which the following expression is undefined.

1. $\dfrac{x-3}{x^2-5x+6}$

1. _____

2. $\dfrac{x+3}{x^2-4}$

2. _____

3. $\dfrac{x-1}{x^2-6x}$

3. _____

Perform the indicated operation.

4. $\dfrac{x^2-3}{x+1}+\dfrac{2x-4}{x+1}$

4. _____

5. $\dfrac{x}{x-5}\cdot\dfrac{2}{x+4}$

5. _____

6. $\dfrac{x-2}{x^2+4x+4}\div\dfrac{2x+1}{x^2-4}$

6. _____

7. $\dfrac{x-3}{x^2-2x-15}-\dfrac{4-x}{x^2-9x+20}$

7. _____

8. $\dfrac{1}{x-1} + \dfrac{x}{1-x}$

8. _____

9. $\dfrac{3(x-4)}{x-1} \cdot \dfrac{x^2-1}{9x}$

9. _____

10. $\dfrac{x-2}{8x-24} \cdot \dfrac{5x-15}{x^2-4}$

10. _____

Simplify each complex fraction.

11. $\dfrac{2-\dfrac{1}{x}}{\dfrac{4}{x^2}-3}$

11. _____

12. $\dfrac{\dfrac{3}{x-1}+\dfrac{4x}{x+1}}{\dfrac{7}{x-1}-\dfrac{2}{x+1}}$

12. _____

Solve each equation.

13. $\dfrac{5}{x-1} = \dfrac{-2}{x+1}$

13. _____

14. $\dfrac{x}{x-1} = \dfrac{1}{2} + \dfrac{3}{x}$

14. _____

15. $\dfrac{3}{n+1} - \dfrac{1}{n+1} = \dfrac{14}{n^2-1}$

15. _____

Solve each equation for the indicated variable.

16. $m = \dfrac{r}{1+rt}$ for r.

16. _____

17. $b = \dfrac{2A - Bh}{h}$ for B

17. _____

Solve.

18. It takes Joan 2 hours to fold the week's worth of laundry. It takes her husband 3 hours. If they work together, how long will it take them to fold the laundry?

18. _____

19. The sum of a number and 5 divided by 6 yields that number less 5, divided by 3. Find the number.

19. _____

Practice Test B

Find any real numbers for which the following expression is undefined.

1. $\dfrac{3x-2}{4x-1}$

    a. 0   b. $\frac{1}{4}$

1. _____

    c. 1   d. $\frac{2}{3}$

2. $\dfrac{x+1}{x^2-1}$

    a. 1 and -1   b. 1

2. _____

    c. 1   d. -1

3. $\dfrac{x+2}{x^2+3}$

    a. -3   b. -2

3. _____

    c. none   d. all real numbers

Perform the indicated operation.

4. $\dfrac{x+1}{x-1}-\dfrac{7}{x+1}$

    a. $\dfrac{x-6}{-2}$ b. $\dfrac{x^2-5x-8}{x^2+1}$

4. _____

    c. $\dfrac{x-6}{2x}$ d. $\dfrac{x^2-5x+8}{x^2-1}$

5. $\dfrac{x^2 + 16x + 64}{x+9} \div \dfrac{x + 17x + 72}{x+8}$

a. $\dfrac{(x+8)^2}{(x+9)^2}$

b. $\dfrac{(x+9)(x+8)}{x+8}$

5. _____

c. $\dfrac{x+9}{(x+8)^2}$

d. $\dfrac{(x+9)^2}{x+8}$

6. $5 - \dfrac{3x}{x-5}$

a. $2x - 5$

b. $\dfrac{2x-25}{x-5}$

6. _____

c. $\dfrac{5x-20}{x-5}$

d. $\dfrac{8x-25}{x-5}$

7. $\dfrac{x-3}{x+1} - \dfrac{4-x}{x}$

a. $\dfrac{2x^2 - 6x - 4}{x(x+1)}$

b. $\dfrac{x^2 - 3x + 4}{x(x+1)}$

7. _____

c. $\dfrac{2x^2 - 6x + 4}{x(x-1)}$

d. $\dfrac{x^2 - 4x - 4}{x(x+1)}$

8. $\dfrac{x+3}{x^2+2x-8} \cdot \dfrac{x-2x}{x^2+6x+9}$

a. $\dfrac{x+3}{x-2}$

b. $\dfrac{(x+3)(x+4)}{(x-2)}$

8. _____

c. $\dfrac{x}{(x+4)(x+3)}$

d. $\dfrac{x(x+3)}{(x-2)(x-3)}$

9. $\dfrac{x+4}{x^2+1} \div \dfrac{x^2-16}{x+1}$

    a. $\dfrac{x+4}{x^2+1}$

    b. $\dfrac{x}{(x+4)(x+3)}$

    c. $\dfrac{x^2+1}{(x+3)(x+4)}$

    d. $\dfrac{x+3}{(x+4)(x^2+1)}$

9. _____

10. $\dfrac{x+1}{x^2+9} + \dfrac{x}{x-3}$

    a. $\dfrac{x^2-9}{(x+1)(x+4)}$

    b. $\dfrac{x^2+1}{(x-3)(x+3)}$

    c. $\dfrac{x^2+4x+1}{x^2-9}$

    d. $\dfrac{(x-3)(x+1)}{x(x+3)}$

10. _____

Simplify each complex fraction.

11. $\dfrac{\dfrac{1}{x}+\dfrac{3}{4x}}{\dfrac{x}{6}-\dfrac{3}{x^2}}$

    a. $\dfrac{2x^3-36}{21x}$

    b. $\dfrac{3x}{2(x-18)}$

    c. $\dfrac{21x}{2x^3-36}$

    d. $\dfrac{18x}{2x^2-36}$

11. _____

12. $\dfrac{\dfrac{3}{x-1}+\dfrac{4x}{x+1}}{\dfrac{7}{x-1}-\dfrac{2}{x+1}}$

    a. $\dfrac{4x^2+7x-3}{9x+9}$

    b. $\dfrac{5x^2-3x+3}{9x+5}$

    c. $\dfrac{7x^2-4x+3}{5x+9}$

    d. $\dfrac{4x^2-x+3}{5x+9}$

12. _____

Solve each equation.

13. $\dfrac{4}{x-5} - \dfrac{3}{x+2} = \dfrac{28}{x^2-3x-10}$

13. _____

   a. 5   b. - 5

   c. -2   d. no solution

14. $\dfrac{x}{x^2-9} + \dfrac{4}{4x-12} = \dfrac{-3}{x}$

14. _____

   a. 1   b. $^{17}\!\!\diagup\!\!_{33}$

   c. $^{33}\!\!\diagup\!\!_{17}$ d. no solution

15. $\dfrac{5-x}{5+x} = \dfrac{6}{7}$

15. _____

   a. -5   b. $^{5}\!\!\diagup\!\!_{13}$

   c. -1 d. no solution

Solve each equation for the indicated variable.

16. $\dfrac{y-b}{x} = m$ for b.

   a. $b = \dfrac{y-m}{x}$ b. $b = \dfrac{y}{mx}$

16. _____

   c. $b = \dfrac{x-m}{y}$ d. $m = y - mx$

Solve.

17. The distance a car travels is directly proportional to the time it is traveling. If a car can travel 200 miles in 4 hours, how long will it take the car to travel 350 miles?

17. _____

   a. 6 hours   b. 5 hours

   c. 8 hours   d. 7 hours

18. Given these are similar triangles, find the measure of x.

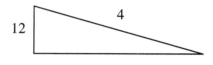

18. _____

   a. 9   b. 1

   c. 6   d. 8

19. It takes a hose 20 hours to fill a pool. Another hose can fill the pool in 15 hours. Approximately how long will it take the pool to fill if both hoses are used?

19. _____

   a. 9.2 hours   b. 8.6 hours

   c. 10.4 hours   d. 9.8 hours

Chapter 8 More on Functions and Graphs
Section 8.1 Graphing and Writing Linear Functions

Learning Objectives
1. Graph linear function
2. Write an equation of a line using function notation.
3. Find equations of parallel and perpendicular lines.

Vocabulary
Use the choices to complete each statement.

Horizontal   **Linear**   **Vertical**
b   **m**   **x**
y

1. The equation $y - 8 = 4$ is a _____ line.

2. A _____ function's equation can be written in the form $f(x) = mx + b$.

3. The equation $x = -5$ is a _____ line.

4. In the equation $y = mx + b$, the _____ represents the slope, and the _____ represents the

    _____ -intercept.

Objective 1

Graph each linear function.

5. $f(x) = -2x + 3$

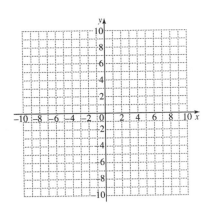

6. $f(x) = -x - 4$

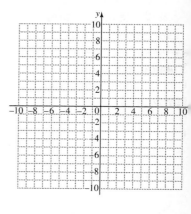

7. $f(x) = \dfrac{2}{3}x + 3$

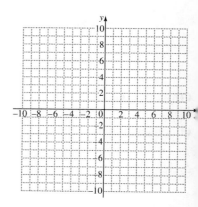

8. $f(x) = 0.25x - 6$

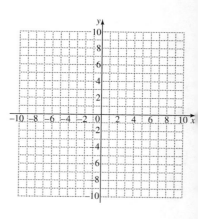

Objective 2

Find the equation of a line with the given slope and containing the given point. Write the equation using function notation.

9. Slope $= \dfrac{1}{4}$; through $(8,-3)$

9. _____

10. Slope $= 3$; through $(1,2)$

10. _____

11. Slope $= -7$; through $\left(\frac{1}{2}, \frac{3}{5} \right)$

11. _____

12. Slope $= -\dfrac{2}{5}$; through $(3,-4)$

12. _____

Objective 3

Find the equation of each line. Write the equation using function notation.

13. Through $(1,-3)$; parallel to $x - 3y = 7$

13. _____

14. Through $(2,-5)$; perpendicular to $3y = x - 6$

14. _____

15. Through $(0,-23)$; perpendicular to $2x - 3y = 8$

15. _____

16. Through $(3,-5)$; parallel to $y = -3$

16. _____

Concept Extension

17. Given the equation $f(x) = 7$, what is the coefficient in front of the x? What is the slope of this line?

17. _____

Find the slope of the line through the following points. Use function notation to write the equation.

18. $(-3,-8),(-6,-9)$

18. _____

19. $\left(\frac{3}{4}, -\frac{2}{5} \right), \left(-\frac{1}{5}, \frac{1}{4} \right)$

19. _____

Section 8.2 Reviewing Function Notation and Graphing Nonlinear Functions

Learning Objectives
1. Review function notation.
2. Find square roots of numbers.
3. Graph nonlinear functions.

Vocabulary
Use the choices to complete each statement.

$y = |x|$ $y = x^2$ $f(9) = 3.6$

$f(3.6) = 9$ 8 -8

6   -6

1. A V-shaped graph could have the equation _____.

2. A parabola could have the equation _____.

3. If the point (9, 3.6) is on the graph of $f(x)$, then _____.

4. If $f(x) = x^3$, then $f(-2) =$ _____.

Objective 1

5. Use the graph of $f(x) =$ to find $f(4)$.   5.  _____

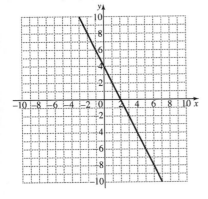

6. Use the graph of $g(x)$ to find $g(-3)$.

6. _____

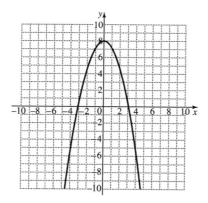

7. Use the graph of $h(x)$ to find $h(4)$.

7. _____

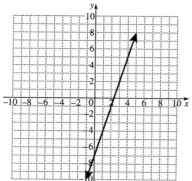

Objective 2

Find the following roots.

 8. $\sqrt{49}$

8. _____

9. $-\sqrt{\dfrac{16}{25}}$

9. _____

 10. $\sqrt{81}$

10. _____

 11. $\sqrt{-100}$

11. _____

12. $\sqrt{-36}$

12. _____

Martin-Gay *Beginning and Intermediate Algebra, Fourth Edition*

Objective 3

Graph the following functions by finding points and plotting ordered pair solutions.

13. $f(x) = 2|x|$

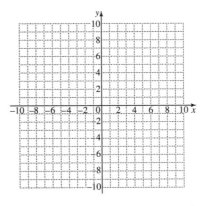

14. $f(x) = x^2 - 4$

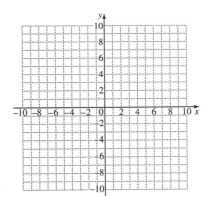

15. $f(x) = x^3 - 7$

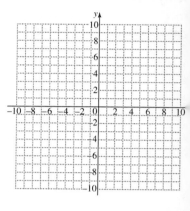

16. $f(x) = |x + 3|$

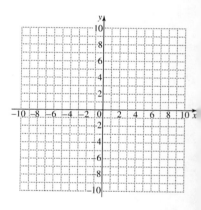

Concept Extension

17. Graph $f(x) = x^2$ and $g(x) = (x-2)^2 + 5$ on the same rectangular coordinate system. Explain what the difference happens to be between the two graphs.

17. _____

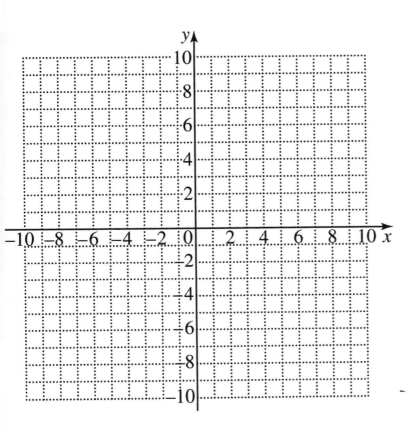

Given the function $f(x) = 3x - 4$.

18. Find $f(a)$ and $f(a-h)$.

18. _____

19. Using the function in number 18, find $\dfrac{f(a) - f(a-h)}{h}$.

19. _____

Section 8.3 Graphing Piecewise-Defined Functions and Shifting and Reflecting Graphs of Functions

Learning Objectives
1. Graph piecewise-defined functions.
2. Vertical and horizontal shifts.
3. Reflect graphs.

Objective 1

Graph each piecewise-defined function.

1. $f(x) = \begin{cases} x+3 & if \ x > 1 \\ -2x+4 & if \ x \le -1 \end{cases}$

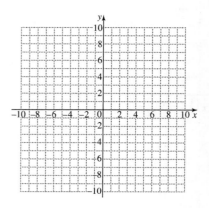

2. $f(x) = \begin{cases} 2x-1 & if \ x > 2 \\ 3x+5 & if \ x \le -2 \end{cases}$

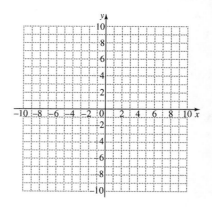

3. $f(x) = \begin{cases} -2 & if \quad x \le -3 \\ 2x-6 & if \quad x > 2 \end{cases}$

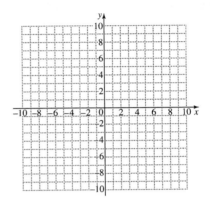

4. $f(x) = \begin{cases} x-5 & if \quad x > 4 \\ 3x-4 & if \quad x \le -1 \end{cases}$

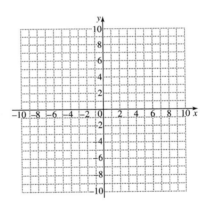

Objective 2

Sketch the graph of each function.

5. $f(x) = x^2 - 4$

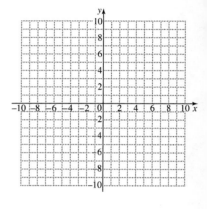

6. $f(x) = |x + 2|$

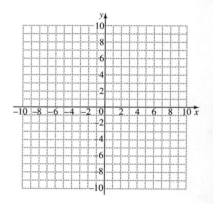

7. $f(x) = (x-1)^2$

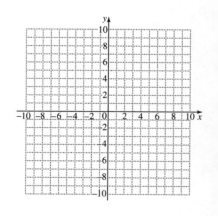

8. $f(x) = \sqrt{x+1} + 1$

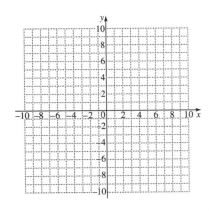

9. $f(x) = |x+2| - 5$

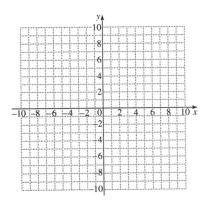

Objective 3

Sketch the graph of each function.

10. $f(x) = -x^2 - 2$

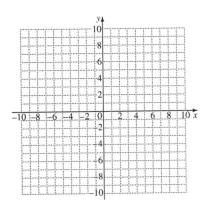

11. $f(x) = -|x+1| + 3$

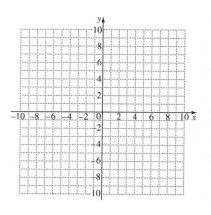

12. $f(x) = -\sqrt{x-2} - 5$

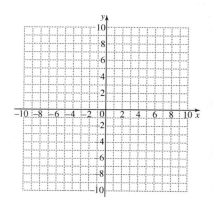

Concept Extension

Graph the following piecewise-defined function.

13. $f(x) = \begin{cases} -2x+6 & \text{if } x > 5 \\ -x+4 & \text{if } 5 < x \leq -2 \\ -3x-9 & \text{if } x < -2 \end{cases}$

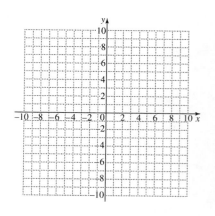

Section 8.4 Variation and Problem Solving
Learning Objective
1.   Solve problems involving direct variation.
2.   Solve problems involving inverse variation.
3.   Solve problems involving joint variation.
4.   Solve problems involving combined variation.

Objective 1

If y varies directly as x, find the constant of variation and the direct variation equation for each situation.

1. $y = 6$ when $x = 42$ 1. _____

2. $y = 10$ when $x = 80$ 2. _____

3. $y = 16$ when $x = 20$ 3. _____

4. $y = 0.6$ when $x = 1.8$ 4. _____

5. $y = \dfrac{3}{4}$ when $x = \dfrac{2}{3}$ 5. _____

Solve.

 6. The weight of a synthetic ball varies directly with the cube of its radius. A ball with a radius of 2 inches weighs 1.20 pounds. Find the weight of a ball of the same material with a 3-inch radius.

6. _____

Objective 2

If y varies inversely as x, find the constant of variation and the inverse variation equation for each situation.

7. $y = 4$ when $x = 6$

7. _____

8. $y = 7$ when $x = 2$

8. _____

9. $y = \dfrac{1}{2}$ when $x = 8$

9. _____

10. $y = 0.3$ when $x = 0.2$

10. _____

Solve.

11. Pairs of markings a set distance apart are made on highways so that the police can detect drivers exceeding the speed limit. Over a fixed distance, the speed R varies inversely with the time T. In one particular pair of markings, R is 45 mph when T is 6 seconds. Find the speed of a car that travels the given distance in 5 seconds.

11. _____

Objective 3

Write each statement as an equation. Use k as the constant of variation.

12. y varies jointly as x and w.

12. _____

13. y varies jointly as the square of x and z; $y = 120$ when $x = 10$ and $z = 2$.

13. _____

14. y varies jointly as the x and z cubed; $y = 320$ when $x = 20$ and $z = 2$.

14. _____

Solve.

15. The maximum weight that a rectangular beam can support varies jointly as its width and the square of its height and inversely as its length. If a beam $\frac{1}{2}$ foot wide, $\frac{1}{3}$ foot high, and 10 feet long can support 12 tons, find how much a similar beam can support if the beam is $\frac{2}{3}$ foot wide, $\frac{1}{2}$ foot high, and 16 feet long.

15. _____

Concept Extension

16. Suppose that y varies directly as x^3. If x is doubled, what is the effect on y?

16. _____

17. Suppose that y varies inversely as x^2. If x is doubled, what is the effect on y?

17. _____

Martin-Gay *Beginning and Intermediate Algebra, Fourth Edition*

Chapter 8 Vocabulary

Vocabulary Word	Definition	Example
$f(x)$	Function notation. A function of x.	$f(x) = 3x + 2$
Linear function	A function that can be written as $f(x) = mx + b$.	$f(x) = 3x - 6$
Vertical Shifts	$g(x) = f(x) + k$; If k positive, graph shifts up; if k is negative, graph shifts down.	$f(x) = x^2 + 3$; graph is three units higher than $f(x) = x^2$
Horizontal Shifts	$g(x) = f(x - h)$; If h is positive, graph shifts to the right; if h is negative, graph shifts to the left.	$f(x) = |x - 4|$; graph is 4 units to the right of $f(x) = |x|$
y varies directly as x.	$y = kx$	$y = 4x$
y varies inversely as x.	$y = \dfrac{k}{x}$	$y = \dfrac{24}{x}$
y varies jointly as x and z.	$y = kxz$	$y = 3xz$

Practice Test A

Use the graph of the function f below.

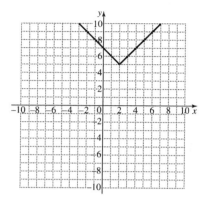

1. Find $f(-2)$.

2. Find all values of x such that $f(x) = 6$.

1. _____

2. _____

Graph each line.

3. $y = 2x - 3$

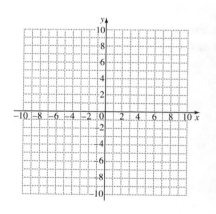

4. $3x - 4y = 6$

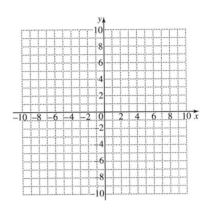

5. $f(x) = \frac{2}{3}x - 5$

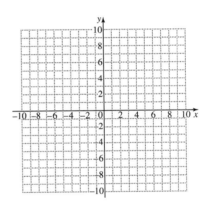

6. $x = -5$

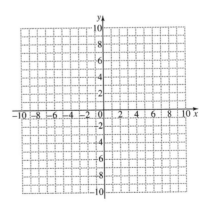

Graph each nonlinear function.

7. $f(x) = (x-3)^2 + 5$

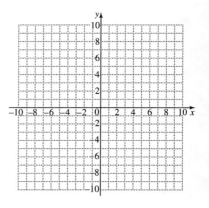

8. $f(x) = |x-3| - 4$

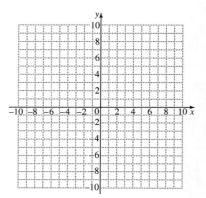

9. $f(x) = -\sqrt{x-2}$

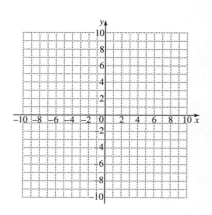

Find the equation of each line satisfying the given conditions. Write the equation using function notation.

10. Perpendicular to $x - 3y = 7$ through $(-6, 2)$.

10. _____

11. Horizontal line through $\left(-\frac{2}{5}, \frac{1}{3}\right)$

11. _____

12. Through $(-2, -4)$ and $(3, 8)$.

12. _____

13. Parallel to $-x + 3y = 6$ through $(-1, -4)$.

13. _____

Determine if the following lines are parallel, perpendicular, or neither.

14. $2y = 3x + 5$
    $6x - 4y = 8$

14. _____

Graph each function.

15. $f(x) = \begin{cases} -2x & if \quad x > 4 \\ x - 4 & if \quad x \le -1 \end{cases}$

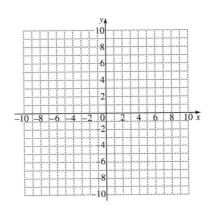

16. $f(x) = \begin{cases} -4 & \text{if} \quad x \geq 3 \\ 2x-6 & \text{if} \quad x < 0 \end{cases}$

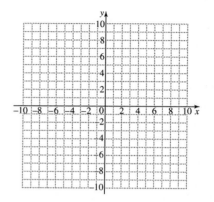

Find the constant of variation and the variation equation.

17. y varies directly with the square of x; $y = 2400$ when $x = 20$.

17. _____

Martin-Gay *Beginning and Intermediate Algebra, Fourth Edition*

Practice Test B

Use the graph of the function $f(x)$ below.

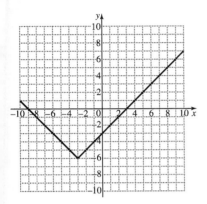

1. Find $f(2)$.

    a. -5 b. -1 c. 5 d. 1

1. _____ .

2. Find all values of x such that $f(x) = -1$.

    a. $-8, 2$ b. -4 c. $-10, 4$ d. -2

2. _____

Find an equation of each line satisfying the given conditions.

3. Horizontal line through the point $(-6, -9)$.

    a. $f(x) = -9$ b. $f(x) = -6$

    c. $x = -9$ d. $x = -6$

3. _____

4. Through the points $(1, 7)$ and $(-4, -8)$.

    a. $f(x) = 3x - 22$ b. $f(x) = 3x + 4$

    c. $f(x) = -3x + 24$ d. $f(x) = -3x - 4$

4. _____

5. Parallel to $2x + 3y = 1$ through $(0, -4)$.

5. _____

    a. $f(x) = -\dfrac{2}{3}x - 4$ b. $f(x) = \dfrac{3}{2}x - 4$

    c. $f(x) = -\dfrac{2}{3}x - \dfrac{4}{3}$ d. $f(x) = -\dfrac{3}{2}x - 2$

Graph the following equations.

6. $-x - y = -2$

6. _____

    a.

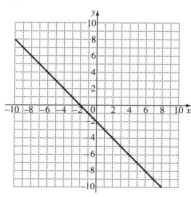

    b.

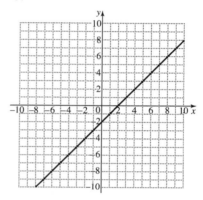

    c.

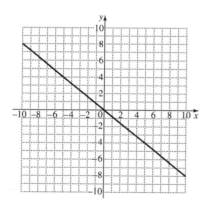

    d.

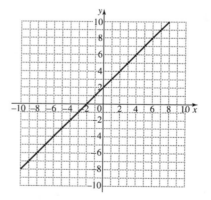

Martin-Gay *Beginning and Intermediate Algebra, Fourth Edition*

7. $f(x) = x^2 - 6$

7. _____

a.

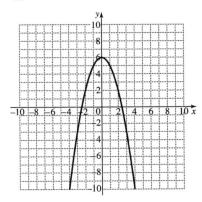

b.

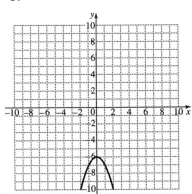

c.

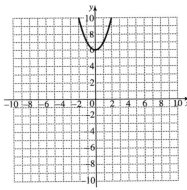

d.

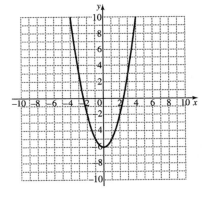

8. $f(x) = |x + 3|$

8. _____

a.

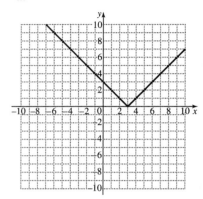

b.

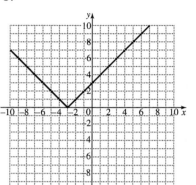

c.

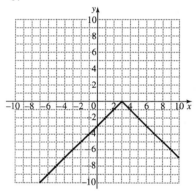

d.

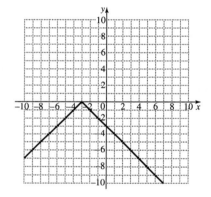

9. $f(x) = -(x+1)^2 - 3$

9. _____

a.

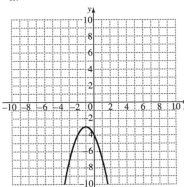

b.

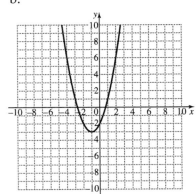

c.

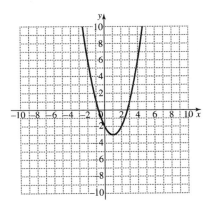

d.

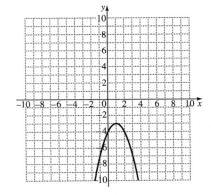

10. $y = -\frac{2}{3}x - 6$

10. _____

a.

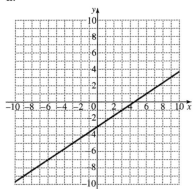

b.

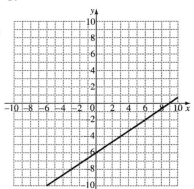

c.

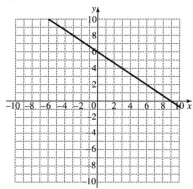

d.

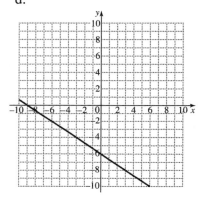

11. $5x - 4y = 12$

11. _____

a.

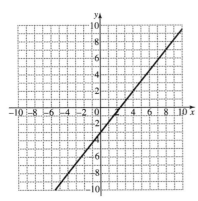

b.

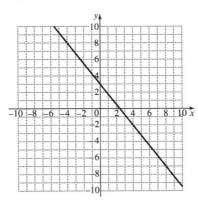

c.

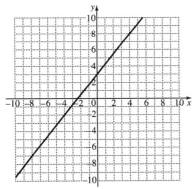

d.

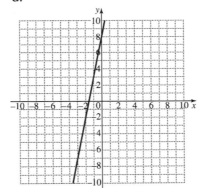

12. Determine whether the following are parallel, perpendicular, or neither.

$$6x + 5y = 18$$
$$5x - 6y = 24$$

12. _____

a. Parallel

b. Perpendicular

c. Neither

Graph each function.

13. $f(x) = \begin{cases} -x-3 & if \ x > 3 \\ 2x+1 & if \ x \le -1 \end{cases}$

13. _____

a.

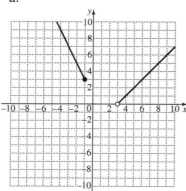

b.

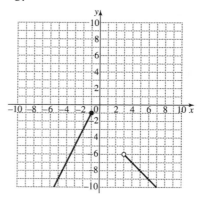

c.

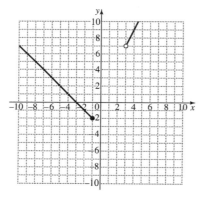

d.

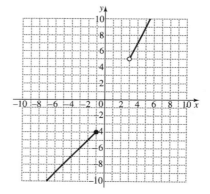

14. $f(x) = \begin{cases} x & \text{if } x > 0 \\ -x & \text{if } x \le 0 \end{cases}$ 14. _____

a.

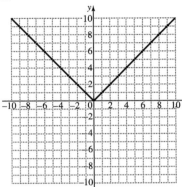

b.

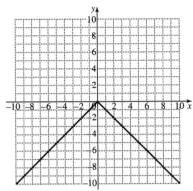

c.

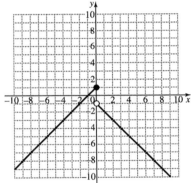

d.

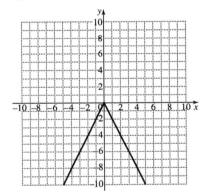

15. Determine whether $y = |x| + 3$ is linear.

                                     15. _____

     a.   Yes it is linear   b.   No, it is not linear.

16. The graph of $y = x^2$ is _____.

     a.   V-shaped   b. a parabola   c. a line

Find the constant of variation and the variation equation.

17. y varies inversely with x; $y = 12$ when $x = -5$.

17. _____

a. $k = -60$; $x = \dfrac{-60}{y}$

b. $k = -\dfrac{12}{5}$; $y = -\dfrac{12}{5}x$

c. $k = -\dfrac{5}{12}$; $y = -\dfrac{5}{12}x$

d. $k = -60$; $y = \dfrac{-60}{x}$

Chapter 9 Inequalities and Absolute Value
Section 9.1 Compound Inequalities

Learning Objectives
1. Find the intersection of two sets.
2. Solve compound inequalities containing **and**.
3. Find the union of two sets.
4. Solve compound inequalities containing **or**.

Vocabulary
Use the choices to complete each statement.

Compound   **Intersection**   **Union**

∩   ∪

1. An inequality that contains two inequalities is called a(n) _____ inequality.

2. The word and means _____ and uses the symbol _____.

3. The word or means _____ and uses the symbol _____.

Objective 1

If $A = \{x \mid x$ is an odd integer between 0 and 30$\}$, $B = \{x \mid x$ is an even number$\}$,
$C = \{2, 5, 7, 13, 18, 22, 27\}$, and $D = \{1, 6, 12, 15, 19, 23, 26, 29\}$, list the elements of each set.

4. $C \cap B$

4. _____

5. $D \cap A$

5. _____

6. $C \cap A$

6. _____

7. $C \cap D$

7. _____

Objective 2

Solve each compound inequality. Write your answer in interval notation.

8. $x \geq 3$ and $x < 10$

8. _____

9. $x \leq 8$ and $x \geq -4$

9. _____

10. $x + 1 \geq 7$ and $3x - 1 \geq 5$

10. _____

11. $-6x \leq -18$ and $x - 20 \leq 10$

11. _____

12. $-4 \leq x + 3 < 11$

12. _____

13. $1 \leq \dfrac{2}{3}x + 3 \leq 4$

13. _____

Objective 3

If $A = \{x \mid x \text{ is an odd integer}\}$, $B = \{x \mid x \text{ is an even number}\}$, $C = \{2, 9, 12, 19\}$, and $D = \{3.8, 13, 18\}$, list the elements of each set.

14. $C \cup D$

14. _____

15. $A \cup C$

15. _____

16. $B \cup D$

16. _____

Objective 4

Solve each compound inequality. Write your answer in interval notation.

17. $x < 2$ or $x < -3$

17. _____

18. $x \geq 5$ or $x \geq -5$

18. _____

19. $2x > 12$ or $x + 7 > 3$

19. _____

20. $4(x - 1) < 9$ or $5(x + 3) \leq 12$

20. _____

21. $x < 1$ or $x > 5$

21. _____

22. $x - 7 > -4$ or $-2x - 14 \geq 20$

22. _____

Concept Extension

Solve each compound inequality. Write your answer in interval notation.

23. $1 - 3x < 4x + 5 \leq x - 5$

23. _____

24. $4(2 + 3x) \leq 2(x + 2) \leq -3(x - 2)$

24. _____

Section 9.2 Absolute Value Equations

Learning Objectives
1. Solve absolute value equations.

Objective 1

Solve each absolute value equation.

1. $|x| = 8$

1. _____

2. $|x + 2| = 10$

2. _____

3. $|2x| - 4 = 16$

3. _____

4. $|z| = -2$

4. _____

5. $|2x - 1| = 5$

5. _____

6. $|x-4|+3=10$

6. _____

7. $\left|\dfrac{x}{2}-3\right|=1$

7. _____

8. $|1+2x|=|3-x|$

8. _____

9. $|x|=|x-4|$

9. _____

10. $|2y-3|=|9-4y|$

10. _____

11. $2 - |3n| = 8$

11. _____

12. $\left| \dfrac{x+1}{8} \right| = \left| \dfrac{x-2}{4} \right|$

12. _____

Concept Extensions

13. Write an absolute value equation representing all numbers x whose distances from -5 is 4 units.

13. _____

14. For what value(s) of c will an absolute value equation of the form $|ax - b| = c$

    A. one solution?
    B. no solution?
    C. two solutions?

14a. _____

14b. _____

14c. _____

Section 9.3 Absolute Value Inequalities

Learning Objectives

1. Solve absolute value inequalities of the form $|x| < a$.

2. Solve absolute value inequalities of the form $|x| > a$.

Objective 1

Solve each inequality. Write your answer in interval notation.

1. $|x| < 5$

1. _____

2. $|3x| \leq 18$

2. _____

3. $|x - 3| < 6$

3. _____

4. $7 - |x + 1| > 4$

4. _____

5. $\left| \dfrac{x+1}{3} \right| \leq 5$

5. _____

6. $|x-5| < -4$

6. _____

7. $|3x-1| < -5$

7. _____

8. $|2x-1| + 4 \leq 9$

8. _____

9. $-15 + |2x-7| \leq -6$

9. _____

Objective 2

Solve each inequality. Write your answer in interval notation.

10. $|x| > -9$

10. _____

11. $|x-1| \geq 5$

11. _____

12. $|1-2x| > 6$

12. _____

13. $\left| \dfrac{x-3}{4} \right| > 6$

13. _____

14. $\left| \dfrac{x+6}{3} \right| > 2$

14. _____

15. $12 - |x-3| \le 7$

15. _____

16. $|2x-3| \ge 7$

16. _____

Section 9.4   Graphing Linear Inequalities in Two Variables and Systems of Linear Inequalities

Learning Objectives
   1. Graph a linear inequality in two variables.
   2. Solve a system of linear inequalities.

Vocabulary.
Use the choices to complete each statement.
True   **False**
Boundary Line   **Linear Inequality in One Variable.**

1. A _____ will divide a plane into two regions called half-planes.

2. $4x + 5y < 12$ is an example of a _____.

3. True or False. The boundary line of the inequality $3x + 2y \geq -5$ is dashed. _____

4. True or False. The shading for the inequality $2x - 3y > 8$ is below the boundary line. _____

5. A _____ of a linear inequality is an ordered pair, that makes the inequality true.

Objective 1

Graph each linear inequality.
6. $x + y < 6$

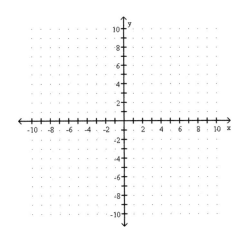

7. $2x + 7y > 5$

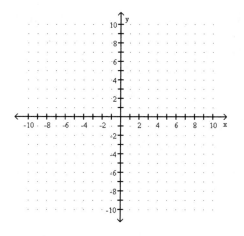

8. $-2x - 4y < -8$

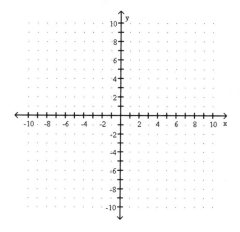

9. $y \geq 2x$

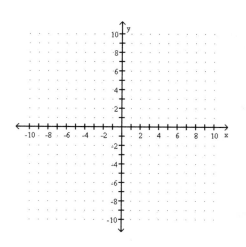

Martin-Gay *Beginning and Intermediate Algebra, Fourth Edition*

10. $x > -5$

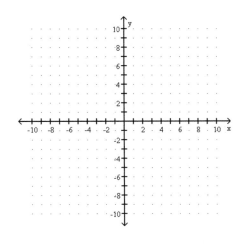

11. $-x < 3y + 9$

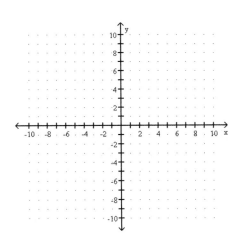

Objective 2

Solve the system of linear inequalities.

12. $\begin{cases} y \le \dfrac{1}{2}x + 3 \\ y > \dfrac{1}{2}x - 1 \end{cases}$

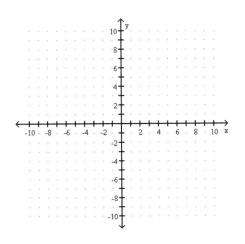

13. $\begin{cases} x \geq 3y \\ x + 3y \leq 6 \end{cases}$

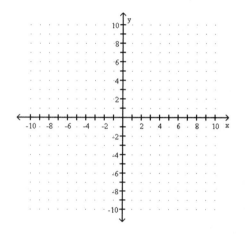

14. $\begin{cases} y \geq 1 \\ x < -3 \end{cases}$

15. $\begin{cases} y < 3x \\ y > -2x \end{cases}$

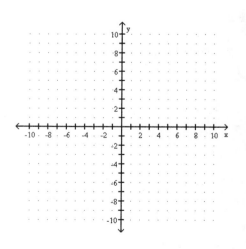

16. $\begin{cases} 4x - 2y > 8 \\ 3x - 2y < 6 \end{cases}$

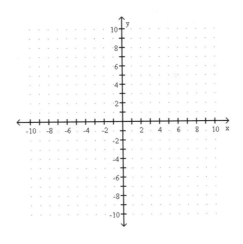

17. $\begin{cases} 6x + 4y < 4 \\ -2x - 4y > 8 \end{cases}$

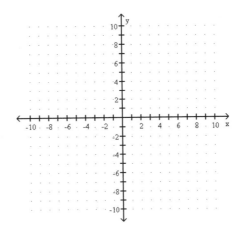

Concept Extensions

18. A local theater holds a maximum of 200 people. For a fundraiser, a youth group will be showing a play. The price for student tickets are $2.00 and the price for nonstudents tickets was set at $3.00 each. The group needs to sell at least $80 worth of ticket to make any profit. Create a graph using only the first quadrant to show all the possible combinations of ticket sales that will allow the youth group to make money.

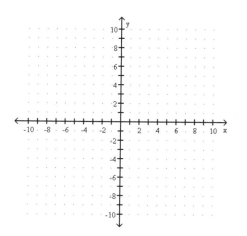

19. Graph this system of linear equations.

$$\begin{cases} x \le 1 \\ y < 7 \\ 5x - y < -8 \\ 3x - 4y < -12 \end{cases}$$

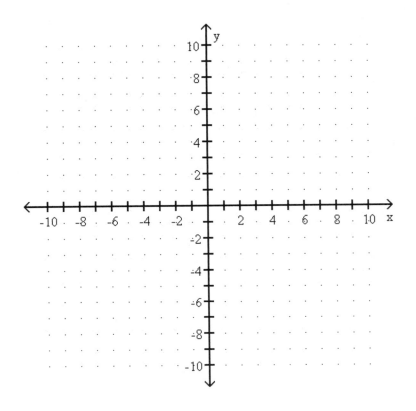

Chapter 9 Vocabulary

Vocabulary Word	Definition	Example
Linear inequality in one variable	An inequality that can be written as $ax + b < c$	$2x - 3 < 6$
Intersection	Inequality that contains and or $\cap$	$x < 2$ and $x > -4$
Union	Inequality that contains or, or $\cup$	$x > 2$ or $x < -4$
Linear Inequality in Two Variables	$Ax + By < C \quad Ax + By > C$ $Ax + By \leq C \quad Ax + By \geq C$	$-3x + 4y < 17$

Practice Test A

Solve each equation or inequality.

1. $|x-2|=5$

1. _____

2. $|x|=-9$

2. _____

3. $|2x-8|=|3x+1|$

3. _____

4. $2x-12=|x-6|$

4. _____

5. $|x-5|\leq 3$

5. _____

6. $x<5$ and $x>-2$

6. _____

7. $|x+8|>5$

7. _____

8. $x<-5$ or $x\geq 2$

8. _____

9. $3 < 2(x - 1) \le 7$

9. _____

10. $-4 \le \dfrac{3x - 5}{2} < 5$

10. _____

11. If $A = \{x \mid x \text{ is an even number}\}$ and $B = \{x \mid 15 < x \le 24\}$, find $A \cap B$.

11. _____

12. Graph the inequality. $x + 4y < 5$

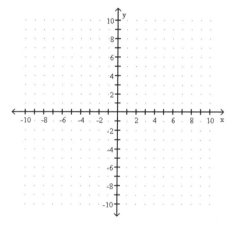

13. Graph the inequality. $2x - 7y > 14$

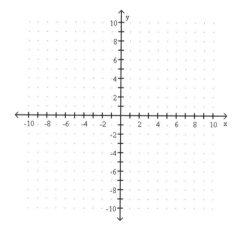

14. Graph the inequality. $-\dfrac{3}{4}x+\dfrac{3}{8}y \le -3$

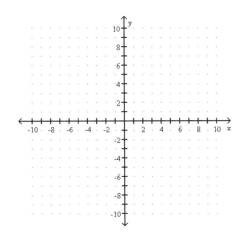

15. Graph the solutions of the system of linear inequality. $\begin{cases} 5x-4y>8 \\ -2x+5y<10 \end{cases}$

16. Graph the solutions of the system of linear inequality. $\begin{cases} x - y \geq 4 \\ x + y < -3 \end{cases}$

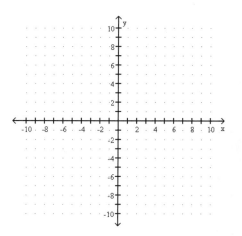

Practice Test B

Solve each equation or inequality.

1. $|x-4|=-3$

      a. 7, 1   b. $-1, -7$

      c. $(-\infty, \infty)$ d. $\varnothing$

1. _____

2. $|2x-6|=|x-5|$

      a. $1, 11/3$ b. $11, 1/3$

      c. $-11, -1/3$ d. $-1, -11/3$

2. _____

3. $|4x-9|=2x+1$

      a. $-5, -4/3$ b. $5, 4/3$

      c. $\varnothing$ d. $(-\infty, \infty)$

3. _____

4. $|x|=9-x$

      a. $9/2$ b. $-9/2$

      c. $(-\infty, \infty)$ d. $\varnothing$

4. _____

5. $17 - |x + 3| = 9$

     a. $-29, 23$ b. $-11, 5$

     c. $\varnothing$ d. $(-\infty, \infty)$

6. $|2x + 7| > 9$

     a. $(-2, 8)$ b. $(-8, 1)$

     c. $(-\infty, -2) \cup (8, \infty)$ d. $(-\infty, -8) \cup (1, \infty)$

7. $x \geq 25$ or $x < 19$

     a. $(19, 25)$ b. $(-\infty, 19) \cup [25, \infty)$

     c. $(19, 25]$ d. $(-\infty, 25) \cup (19, \infty)$

8. $|3x - 4| < 5$

     a. $\left(-\dfrac{1}{3}, 3\right)$ b. $\left(-\infty, -\dfrac{1}{3}\right) \cup (3, \infty)$

     c. $\varnothing$ d. $\left(-3, \dfrac{1}{3}\right)$

9. $x \leq -6$ and $x < 4$

     a. $[-6, 4)$ b. $(-\infty, 4)$

     c. $(-\infty, -6]$ d. $(-\infty, -6)$

Martin-Gay *Beginning and Intermediate Algebra, Fourth Edition*

10. $-1 \leq \dfrac{2x-7}{3} < 5$

10. _____

    a. (−2, 11)   b. [2, 11)

    c. [−2, 11)   d. (2, 11)

11. Find the union of $A = \{2,5,8,9\}$ and $B = \{1,6,8,11\}$.

11. _____

    a. $\{5,6,8,9\}$ b. $\{8\}$

    c. $\varnothing$ d. $\{1,2,5,6,8,9,11\}$

12. Graph the inequality. $2x - 5y > -10$

12. _____

  a.

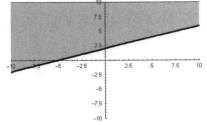

  b.

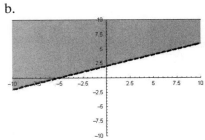

  c.

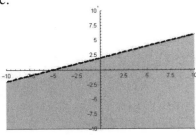

  d.

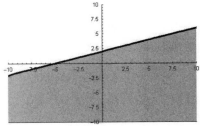

13. Graph the inequality. $x \leq -2y - 6$

13. _____

a.

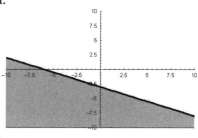

b.

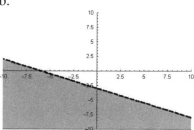

c.

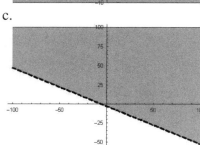

d.

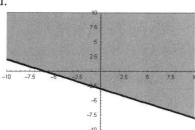

14. Graph the inequality. $0.25x - 0.60y \geq -0.15$

14. _____

a.

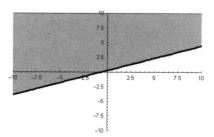

b.

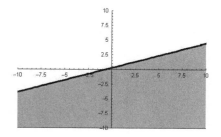

c.

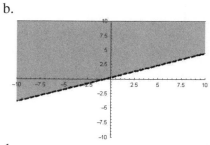

d.

15. Graph the solutions of the system of linear inequalities. $\begin{cases} y < \dfrac{3}{4}x - 2 \\ 3x - 4y \le 12 \end{cases}$

a.

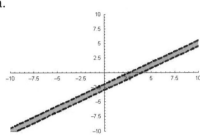

b.

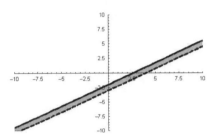

c.

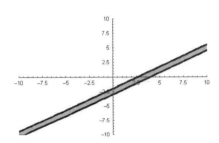

d.

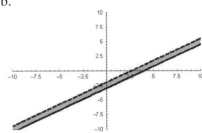

15. _____

16. Graph the solutions of the system of linear inequalities. $\begin{cases} x + 3y > 5 \\ -x - 5y < -8 \end{cases}$

a.

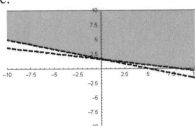

b.

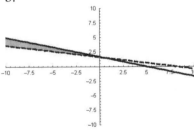

c.

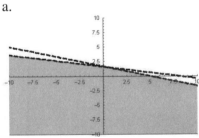

d.

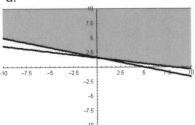

16. _____

Chapter 10 Rational Exponents, Radicals, and Complex Numbers
Section 10.1 Radicals and Radical Functions

Learning Objectives
 1. Find square roots.
 2. Approximate roots.
 3. Find cube roots.
 4. Find nth roots.
 5. Find $\sqrt[n]{a^n}$ where a is a real number.
 6. Graph square and cube root functions.

Vocabulary.
Use the choices to complete each statement.

Index	**Radical sign**	**Radicand**
Cubes	**Squares**	**True**
False		

1. In the expression $\sqrt[4]{5}$, 4 is the _____; 5 is the _____, and the $\sqrt{}$ is

   the _____.

2. True or false. The square root of -4 is -2. _____

3. The numbers 9, 25, and 81 are perfect _____.

4. The numbers 1, 27, and 64 are perfect _____.

Objective 1

Simplify.

5. $\sqrt{3600}$

5. _____

6. $\sqrt{\dfrac{4}{9}}$

6. _____

7. $\sqrt{0.25}$

7. _____

Martin-Gay *Beginning and Intermediate Algebra, Fourth Edition*

8. $-\sqrt{49}$ 8. _____

9. $\sqrt{-16}$ 9. _____

10. $\sqrt{81x^{16}}$ 10. _____

Objective 2

Use a calculator to approximate each square root to 3 decimal places.

11. $\sqrt{24}$ 11. _____

12. $\sqrt{120}$ 12. _____

13. $\sqrt{104}$ 13. _____

14. $\sqrt{260}$ 14. _____

Objective 3

Find each cube root.

15. $\sqrt[3]{64}$ 15. _____

16. $\sqrt[3]{\dfrac{1}{-27}}$

16. _____

 17. $\sqrt[3]{-27x^9}$

17. _____

18. $\sqrt[3]{1000}$

18. _____

Objective 4

Find each root. Assume all variables represent nonnegative real numbers.

19. $\sqrt[5]{243}$

19. _____

 20. $\sqrt[6]{64x^{12}}$

20. _____

21. $\sqrt[4]{-81x^{16}}$

21. _____

Objective 5

Simplify, Assume all variables represent nonnegative real numbers.

22. $\sqrt{(x-5)^2}$

22. _____

23. $\sqrt[8]{y^8}$ 23. _____

24. $\sqrt{(-5)^2}$ 24. _____

25. $\sqrt[3]{(x-1)^3}$ 25. _____

Objective 6

Graph the following functions.

26. $f(x) = \sqrt{x+5}$

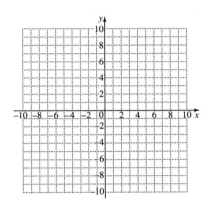

27. $g(x) = \sqrt[3]{x-2}$

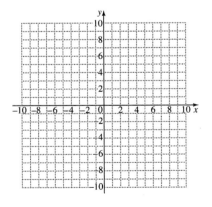

28. $h(x) = \sqrt{x} - 4$

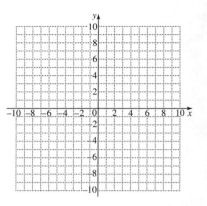

Concept Extension

Without a calculator, state what whole number the following square roots are closest to.

29. $\sqrt{389}$

29. _____

30. $\sqrt{2480}$

30. _____

Section 10.2 Rational Exponents

Learning Objectives

1. Understand the meaning of $a^{1/n}$.
2. Understand the meaning of $a^{m/n}$.
3. Understand the meaning of $a^{-m/n}$
4. Use rules for exponents to simplify expressions that contain rational exponents.
5. Use rational exponents to simplify radical expressions.

Objective 1

Use radical notation to write each expression. Simplify if possible.

1. $25^{1/2}$ 1. _____

2. $-(32)^{1/5}$ 2. _____

3. $(5x)^{1/3}$ 3. _____

4. $4x^{1/4}$ 4. _____

Objective 2

Use radical notation to write each expression. Simplify if possible.

5. $36^{3/2}$ 5. _____

6. $(x+3)^{2/5}$ 6. _____

 7. $(-64)^{2/3}$ 7. _____

8. $(4x+3)^{5/6}$ 8. _____

Objective 3

Write with positive exponents. Simplify if possible.

9. $4^{-1/2}$ 9. _____

10. $x^{-1/5}$ 10. _____

11. $\dfrac{4}{5x^{-3/4}}$ 11. _____

12. $(-32)^{-5/6}$ 12. _____

Objective 4

Use the properties of exponents to simplify each expression. Write with positive exponents.

13. $x^{1/2}x^{1/4}$ 13. _____

14. $\left(16y^4\right)^{3/4}$

14. _____

15. $\dfrac{5^{2/3}5^{1/4}}{5^{1/3}}$

15. _____

16. $\dfrac{\left(3x^{1/4}\right)^3}{x^{1/12}}$

16. _____

Multiply.

17. $\left(x^{1/2}-2\right)\left(x^{1/3}+1\right)$

17. _____

18. $x^{2/3}\left(x-2\right)$

18. _____

Factor the common factor from the given expression.

19. $x^{3/5}; x^{4/5}-3x$

19. _____

20. $x^{1/3}; x^{4/3}-4x^{5/3}$

20. _____

Objective 5

Use rational exponents to simplify each radical. Assume that all variables represent positive numbers.

21. $\sqrt[4]{y^2}$

21. _____

22. $\sqrt[15]{r^5 s^{10}}$

22. _____

23. $\sqrt[6]{x^{21} y^{15}}$

23. _____

Use rational exponents to write as a single expression.

24. $\sqrt[3]{x^2} \cdot \sqrt[4]{x^3}$

24. _____

25. $\sqrt{5} \cdot \sqrt[3]{6}$

25. _____

26. $\dfrac{\sqrt[3]{x^2}}{\sqrt[5]{x}}$

26. _____

Concept Extension

27. What times $x^{2/3}$ will the result be $x^{1/4}$?

27. _____

28. What divided by $x^{5/8}$ will the result be $x^{7/4}$?

28. _____

Section 10.3 Simplifying Radical Expressions

Learning Objectives
1. Use the product rule for radicals.
2. Use the quotient rule for radicals.
3. Simplify radicals.
4. Use the distance and midpoint formula.

Objective 1

Use the product rule to multiply.

1. $\sqrt{5} \cdot \sqrt{7}$

1. _____

2. $\sqrt{13} \cdot \sqrt{x}$

2. _____

3. $\sqrt{\dfrac{5}{6}} \cdot \sqrt{11}$

3. _____

4. $\sqrt[5]{13ab} \cdot \sqrt[5]{2a^2b^3}$

4. _____

5. $\sqrt[3]{\dfrac{3}{4}} \cdot \sqrt[3]{\dfrac{2}{5}}$

5. _____

Objective 2

Use the quotient rule to simplify.

6. $\sqrt{\dfrac{3}{16}}$

6. _____

7. $\sqrt{\dfrac{81}{x^2z}}$

7. _____

8. $\sqrt[3]{\dfrac{27y}{8x^6}}$

8. _____

9. $\sqrt{\dfrac{z^{16}}{25x^{12}}}$

9. _____

Objective 3

Simplify.

10. $\sqrt{50}$

10. _____

11. $2\sqrt{8x^2}$

11. _____

12. $\sqrt[3]{250x^4}$

12. _____

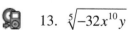 13. $\sqrt[5]{-32x^{10}y}$

13. _____

14. $-\sqrt[4]{32x^6y^5}$

14. _____

Objective 4

Find the distance between each pair of points. Give an exact distance and a three-decimal-place approximation.

15. $(1,2)$ and $(-3,5)$ 15. _____

16. $(-3,2)$ and $(1,-3)$ 16. _____

17. $(0,4)$ and $(-2,-6)$ 17. _____

Find the midpoint of the line segment whose endpoints are given.

18. $(2,3)$ and $(-4,5)$ 18. _____

19. $(-3,-6)$ and $(6,4)$ 19. _____

20. $(-2,-1)$ and $(-8,6)$ 20. _____

Concept Extension

Find the midpoint and distance between the given points.

21. $\left(\sqrt{3},\sqrt{5}\right)$ and $\left(-3\sqrt{3},6\sqrt{5}\right)$ 21. _____

Section 10.4 Adding, Subtracting, and Multiplying Radical Expressions

Learning Objectives
1. Add or subtract radical expressions.
2. Multiply radical expressions.

Objective 1

Add or subtract.

 1. $2\sqrt{50} - 3\sqrt{125} + \sqrt{98}$

1. _____

2. $3\sqrt[3]{16x^3} - x\sqrt[3]{2} + \sqrt[3]{54x}$

2. _____

3. $3\sqrt{y^3} + \sqrt{y^5} + \sqrt{y}$

3. _____

4. $\sqrt[3]{\dfrac{11}{8}} - \dfrac{\sqrt[3]{11}}{6}$

4. _____

5. $\sqrt{\dfrac{12x}{25}} - \dfrac{\sqrt{48x}}{5}$

5. _____

Objective 2

Multiply, and then simplify if possible.

6. $\sqrt{3}\left(2 + \sqrt{3}\right)$

6. _____

7. $\sqrt{6}\left(\sqrt{12} - \sqrt{24}\right)$

7. _____

8. $\left(2 + \sqrt{3}\right)\left(4 - \sqrt{3}\right)$

8. _____

9. $\left(2\sqrt{7} + 3\sqrt{5}\right)\left(\sqrt{7} - 2\sqrt{5}\right)$

9. _____

10. $\left(\sqrt[3]{4} - 5\right)\left(3 - \sqrt[3]{2}\right)$

10. _____

11. $\left(\sqrt{3}-2\right)^{2}$ 11. _____

12. $\left(\sqrt{y+1}-3\right)^{2}$ 12. _____

Concept Extension

13. Find the perimeter of a rectangle with the length of $\sqrt{108}$ feet and width of $\sqrt{75}$ feet.

13. _____

14. Find the length of the diagonal for the rectangle above. Round the answer to three decimal places, if necessary.

14. _____

Section 10.5 Rationalizing Denominators and Numerators of Radical Expressions

Learning Objectives
1. Rationalize denominators.
2. Rationalize denominators having two terms.
3. Rationalize numerators.

Vocabulary.
Use the choices to complete each statement.

Conjugate   **Rationalizing the Numerator**   **Rationalizing the Denominator**

$\dfrac{6}{6}$ $\dfrac{\sqrt{5}}{\sqrt{5}}$

1. _____ is the process of removing radicals from the numerator.

2. The _____ of $5+\sqrt{2}$ is $5-\sqrt{2}$.

3. To rationalize the denominator of $\dfrac{6}{\sqrt{5}}$, you will multiply the expression by _____.

Objective 1

Rationalize each denominator.

4. $\dfrac{4}{\sqrt{3}}$ 4. _____

5. $\sqrt{\dfrac{5}{6x}}$ 5. _____

6. $\dfrac{3}{\sqrt{8x}}$ 6. _____

7. $\dfrac{8x}{\sqrt[5]{8x^4}}$

7. _____

Objective 2

Rationalize each denominator.

8. $\dfrac{3}{2+\sqrt{x}}$

8. _____

9. $\dfrac{\sqrt{3}+\sqrt{5}}{4-\sqrt{2}}$

9. _____

10. $\dfrac{3\sqrt{x}+y}{2\sqrt{x}-\sqrt{y}}$

10. _____

11. $\dfrac{-7}{\sqrt{x}-3}$

11. _____

Objective 3

Rationalize each numerator.

12. $\sqrt{\dfrac{18}{5}}$

12. _____

13. $\dfrac{\sqrt{3x}}{15}$

13. _____

14. $\sqrt[3]{\dfrac{54}{243}}$

14. _____

15. $\dfrac{1+\sqrt{x}}{3}$

15. _____

16. $\dfrac{\sqrt{a}+\sqrt{b}}{\sqrt{a}-\sqrt{b}}$

16. _____

17. $\dfrac{\sqrt{2} - 4x}{x + \sqrt{2}}$

17. _____

Concept Extension

18. The formula for the radius r of a sphere with a volume of V is $r = \sqrt[3]{\dfrac{3V}{4\pi}}$. Rationalize the denominator of the radical expression in this formula.

18. _____

Section 10.6 Radical Equations and Problem Solving

Learning Objectives
1. Solve equations that contain radical expressions.
2. Use the Pythagorean Theorem to model problems.

Objective 1

Solve.

1. $\sqrt{x+1} = 6$ 1. _____

2. $\sqrt{2x+5} + 2 = 7$ 2. _____

3. $\sqrt{x-3} = -6$ 3. _____

4. $\sqrt[3]{x-7} = 3$ 4. _____

5. $\sqrt[3]{2x-3} - 2 = -5$

5. _____

6. $\sqrt{x+8} = \sqrt{2x+6}$

6. _____

7. $x + \sqrt{x+1} = 1$

7. _____

8. $\sqrt{x+2} - \sqrt{x-2} = 1$

8. _____

9. $\sqrt{5x-1} - \sqrt{x} + 2 = 3$

9. _____

Objective 2

Solve.

Find the length of the unknown side of the triangle.

10.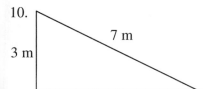

10. _____

11. A two cars leave Dallas, one heads north, the other heads east. The car traveling north averages a
     speed of 50 mph while the other car averages 45 mph. After 3 hours, how far apart are the two cars?
     Round your answer to the nearest whole number.

11. _____

12. Find the length of the unknown side.

6 $5\sqrt{2}$

12. _____

Concept Extension

13. $\sqrt{(x^2 - x) + 4} = 3(x^2 - x) + 2$ 13. _____

14. Solve: $\sqrt{\sqrt{x - 2} + \sqrt{x}} = \sqrt{5}$

14. _____

Section 10.7 Complex Numbers

Learning Objectives
1. Write square roots of negative numbers in the form bi.
2. Add or subtract complex numbers.
3. Multiply complex numbers.
4. Divide complex numbers.
5. Raise i to powers.

Vocabulary
Use the choices to complete each statement.

Complex	Imaginary Unit	Pure Imaginary
Real	- 1	1
$\sqrt{-1}$	$\sqrt{1}$	

1. In the complex number system, i is the _____.

2. A _____ number is any number that can be written in the form: $a + bi$, where both a and b are real numbers.

3. A _____ is the complex number, where $a = 0$ and b does not.

4. $i =$ _____.

5. A complex number is considered a _____ number, if $b = 0$.

6. $i^2 =$ _____.

Objective 1

Simplify.

7. $\sqrt{-36}$

7. _____

8. $\sqrt{-25}$

8. _____

9. $\sqrt{-5}$ 9. _____

10. $\sqrt{-72}$ 10. _____

Objective 2

Add or subtract. Write the sum or difference in the form $a + bi$.

11. $(4 + i) + (3 - 2i)$ 11. _____

12. $(5 + 6i) - (4 + 2i)$ 12. _____

13. $(5 + 4i) + (3 + 2i) - (4 - 3i)$ 13. _____

14. $(4 + 10i) - (11 + 2i)$ 14. _____

Objective 3

Multiply. Write the product in the form $a + bi$.

15. $-5i \cdot 12i$ 15. _____

16. $4i \cdot (2 + 3i)$

16. _____

17. $(3 - 7i)(2 - 3i)$

17. _____

18. $(8 - 2i)^2$

18. _____

19. $(6 - 2i)(3 + i)$

19. _____

Objective 4

Write each quotient in the form $a + bi$.

20. $\dfrac{3}{2i}$

20. _____

21. $\dfrac{4}{2 - i}$

21. _____

22. $\dfrac{3+2i}{1+5i}$ 22. _____

23. $\dfrac{3+5i}{1+i}$ 23. _____

Objective 5

Find each power of i.

24. i^9 24. _____

25. i^{30} 25. _____

26. $(2i)^6$ 26. _____

27. $(-4i)^3$ 27. _____

Concept Extension

Write in the form $a + bi$.

28. $\dfrac{4 - \sqrt{-4}}{6}$

28. _____

29. $\dfrac{7 - \sqrt{-72}}{14}$

29. _____

Chapter 10 Vocabulary

Vocabulary Word	Definition	Example
Principal, or square root	$\sqrt{a} = b$ if $b^2 = a$	$\sqrt{100} = 10$
Cube root	$\sqrt[3]{a} = b$ if $b^3 = a$	$\sqrt[3]{27} = 3$
Radical Function	Function that contains a root of x.	$f(x) = \sqrt{x} - 4$
Distance Formula	$D = \sqrt{(x_2 - x_1)^2 + (y_2 - y_1)^2}$	$(0,2)(-2,4)$ $\sqrt{(1+2)^2 + (2-4)^2} = \sqrt{9+4} = \sqrt{1}$
Midpoint Formula	$\left(\dfrac{x_1 + x_2}{2}, \dfrac{y_1 + y_2}{2} \right)$	$(0,2)(-2,4)$ $\left(\dfrac{0-2}{2}, \dfrac{2+4}{2} \right) = \left(\dfrac{-2}{2}, \dfrac{6}{2} \right) = (-1,3)$
Like Radical	Radicals with the same index and the same radicand	$5\sqrt{3}$ and $x\sqrt{3}$ are like terms
Conjugate	The conjugate of $a + b$ is $a - b$	$4c + 2$ and $4c - 2$ are conjugates
Rationalizing the denominator	The process of writing the denominator without the radicals.	$\dfrac{2}{\sqrt{3}} \cdot \dfrac{\sqrt{3}}{\sqrt{3}} = \dfrac{2\sqrt{3}}{3}$
Rationalizing the numerator	The process of writing the denominator without the radicals.	$\dfrac{\sqrt{2}}{5} \cdot \dfrac{\sqrt{2}}{\sqrt{2}} = \dfrac{2}{5\sqrt{2}}$
Complex number	Any number that can be written as $a + bi$	$4 + 3i$
Complex Conjugates	$a + bi$ and $a - bi$ are conjugates.	The conjugate of $2 + 3i$ is $2 - 3i$

Practice Test A

Raise to the power or find the root. Assume that all variables represent positive numbers. Write with only positive exponents.

1. $\sqrt{320}$

1. _____

2. $-\sqrt[5]{-32x^4}$

2. _____

3. $\left(\dfrac{8}{27}\right)^{-\frac{1}{3}}$

3. _____

4. $x^{\frac{1}{2}}\left(x^{\frac{2}{3}} - x^{\frac{1}{4}}\right)$

4. _____

5. $\left(a^4 b^6\right)^{\frac{3}{2}}$

5. _____

6. $\left(\dfrac{49x^{\frac{2}{3}}}{y^{\frac{4}{5}}}\right)^{\frac{1}{2}}$

6. _____

7. $\sqrt{-27a^2}$

7. _____

8. $\sqrt{-64}$

8. _____

Rationalize the denominator.

9. $\sqrt{\dfrac{16}{5}}$

9. _____

10. $\dfrac{5+\sqrt{x}}{3-\sqrt{x}}$

10. _____

11. $\dfrac{\sqrt[3]{xy}}{\sqrt[3]{x^2 y}}$

11. _____

12. $\dfrac{4+3i}{6-i}$

12. _____

Perform the indicated operation.

13. $\left(8+\sqrt{2}\right)\left(4-\sqrt{2}\right)$

13. _____

14. $\sqrt{5}\left(\sqrt{3}-\sqrt{2}\right)$

14. _____

15. $\cdot\left(\sqrt{7}-x\right)^{2}$

15. _____

16. $3\sqrt{72x^{2}}-x\sqrt{50}+\sqrt{8x}$

16. _____

Use a calculator to approximate each to three decimal places.

17. $\sqrt{172}$

17. _____

18. $452^{1/4}$

18. _____

Solve.

19. $\sqrt{2x+5}=7$

19. _____

20. $\sqrt{x^2 + 3x} = \sqrt{8x + 6}$

20. _____

21. $x - \sqrt{x + 2} = 4$

21. _____

Perform the indicated operation and simplify.

22. $(3 + 4i)(5 - 3i)$

22. _____

23. $(4 + i) - (3 + 4i)$

23. _____

24. Find the distance between $(-4, 5)$ and $(1, -6)$.

24. _____

25. Find the midpoint of the line segment whose endpoints are $(-2, -4)$ and $(7, 8)$.

25. _____

Practice Test B

Raise to the power or find the root. Assume that all variables represent positive numbers. Write with only positive exponents.

1. $\sqrt{140}$

1. _____

    a. $2\sqrt{35}$ b. $4\sqrt{35}$

    c. $5\sqrt{14}$ d. $10\sqrt{7}$

2. $-\sqrt[3]{-64y^6}$

2. _____

    a. $8y^3$ b. $4y^2$

    c. $-8y^2$ d. Not a real number

3. $\left(\dfrac{16}{81}\right)^{-3/4}$

3. _____

    a. $-\dfrac{8}{27}$ b. $\dfrac{8}{27}$

    c. $\dfrac{4}{9}$ d. $\dfrac{27}{8}$

4. $x^{1/4}\left(x^2 + x^{1/2}\right)$

4. _____

    a. $x^{1/2} + x^{1/8}$ b. $x^{9/4} + x^{3/4}$

    c. $x^{1/2} + x^{3/4}$ d. $x^{9/4} + x^{1/8}$

5. $\left(x^6 y^9\right)^{2/3}$

    a. $x^4 y^5$

    c. $x^4 y^6$

    b. $x^{20/3} y^{29/3}$

    d. $x^9 y^{12}$

5. _____

6. $\left(\dfrac{125 a^{3/4}}{b^2}\right)^{-1/3}$

    a. $\dfrac{b^{2/3}}{5 a^{1/4}}$

    c. $\dfrac{5 a^{5/12}}{b^{5/3}}$

    b. $\dfrac{5 a^{1/4}}{b^{2/3}}$

    d. $-\dfrac{5 a^{1/4}}{b^{2/3}}$

6. _____

7. $\sqrt{-81 x^6}$

    a. $9 x^3 i$

    c. $-9 x^3$

    b. $-9 x^3$

    d. $9 x^4 i$

7. _____

8. $\sqrt{-125}$

    a. $-5\sqrt{5}$

    c. $-5i$

    b. $5\sqrt{5} i$

    d. $25i$

8. _____

Rationalize the denominator.

9. $\sqrt{\dfrac{25}{6}}$

9. _____

    a. $\dfrac{5\sqrt{6}}{6}$

    b. $\dfrac{5}{6}$

    c. $\dfrac{5\sqrt{3}}{2}$

    d. $\dfrac{5}{2}$

10. $\dfrac{6-\sqrt{3}}{5+\sqrt{2}}$

10. _____

    a. $\dfrac{\sqrt{6}+5\sqrt{3}-6\sqrt{2}-30}{29}$

    b. $\dfrac{\sqrt{6}+5\sqrt{3}-6\sqrt{2}-30}{23}$

    c. $\dfrac{\sqrt{6}-5\sqrt{3}-6\sqrt{2}+30}{29}$

    d. $\dfrac{\sqrt{6}-5\sqrt{3}-6\sqrt{2}+30}{23}$

11. $\dfrac{\sqrt[4]{ab^2}}{\sqrt[4]{a^2b}}$

11. _____

    a. $\dfrac{\sqrt[4]{a^3b}}{a}$

    b. $\dfrac{\sqrt[4]{ab^3}}{ab}$

    c. $\dfrac{\sqrt[4]{a^3b^2}}{a}$

    d. $\dfrac{\sqrt[4]{a^3b^3}}{ab}$

12. $\dfrac{5-2i}{8+2i}$

12. _____

     a. $\dfrac{5}{8} - i$

     b. $\dfrac{5}{4} - \dfrac{1}{2}i$

     c. $\dfrac{9}{17} - \dfrac{13}{34}i$

     d. $\dfrac{11}{17} - \dfrac{3}{24}i$

Perform the indicated operation.

13. $\left(7 - \sqrt{3}\right)\left(8 - \sqrt{3}\right)$

13. _____

     a. $59 - 15\sqrt{3}$

     b. $53 - \sqrt{3}$

     c. $53 + \sqrt{3}$

     d. $15\sqrt{3} + 59$

14. $\sqrt{10}\left(\sqrt{12} + \sqrt{15}\right)$

14. _____

     a. $4\sqrt{30} + 25\sqrt{6}$

     b. $2\sqrt{30} + 5\sqrt{6}$

     c. $2\sqrt{30} - 5\sqrt{6}$

     d. $\sqrt{120} + \sqrt{50}$

15. $\left(\sqrt{2x} - 5\right)^2$

15. _____

     a. $2x - 10\sqrt{2x} + 25$

     b. $4x + 10\sqrt{2x} - 25$

     c. $27 - 5\sqrt{2x}$

     d. $27x - 10\sqrt{2x}$

16. $5\sqrt{108} + x\sqrt{75} - \sqrt{27}$ 16. _____

    a. $32x\sqrt{3}$ b. $27\sqrt{3} + 5x$

    c. $5x\sqrt{3} + 3$ d. $(5x + 27)\sqrt{3}$

Use a calculator to approximate each to three decimal places.

17. $\sqrt{642}$ 17. _____

    a. 25.338   b. 24.447

    c. 25.663   d. 26.116

18. $225^{\frac{1}{3}}$ 18. _____

    a. 6.893   b. 7.042

    c. 6.082   d. 6.889

Solve.

19. $\sqrt{4x - 7} = 5$ 19. _____

    a. 8   b. 3.5

    c. 3   d. $\varnothing$

20. $\sqrt{2x + 16} = \sqrt{5x + 7}$ 20. _____

    a. 4   b. -3

    c. 3   d. $\varnothing$

21. $2x + \sqrt{x-1} = 3$

21. _____

    a. $\dfrac{5}{4}$ b. 1

    c. 2   d. $\varnothing$

Perform the indicated operation and simplify.

22. $(4 - 5i)(2 + 6i)$

22. _____

    a. $-22 + 34i$ b. $38 - 14i$

    c. $-22 - 34i$ d. $38 + 14i$

23. $(6 + 2i) - (7 + 5i)$

23. _____

    a. $-1 + 7i$ b. $-1 - 3i$

    c. $13 - 3i$ d. $13 + 7i$

24. Find the distance between $(-2, -6)$ and $(4, 8)$.

24. _____

    a. $2\sqrt{2}$ b. $4\sqrt{29}$

    c. $2\sqrt{58}$ d. $4\sqrt{10}$

25. Find the midpoint of the line segment whose endpoints are $(1,7)$ and $(-4,-8)$.

25. _____

a. $\left(\frac{7}{2}, \frac{11}{2}\right)$

b. $\left(\frac{5}{2}, \frac{15}{2}\right)$

c. $(3,7)$

d. $\left(-\frac{3}{2}, -\frac{1}{2}\right)$

Chapter 11 Quadratic Equations and Functions
Section 11.1 Solving Quadratic Equations by Completing the Square

Learning Objective
1. Use the square root property to solve quadratic equations.
2. Solve quadratic equations by completing the square.
3. Use quadratic equations to solve problems.

Vocabulary
Use the choices to complete each statement.

Completing the square	Quadratic Equation	$\sqrt{b}$
$-\sqrt{b}$	$\pm\sqrt{b}$	b^2
$\dfrac{b}{2}$	$\left(\dfrac{b}{2}\right)^2$	

1. A _____ is an equation that can be written in the form $ax^2 + bx + c = 0$, $a \neq 0$.

2. According to the square root property, if $a^2 = b$ then $a =$ _____, if b is a real number.

3. One method to solve a quadratic equation is _____, this method is the process of rewriting one side to be a perfect square trinomial.

4. To solve a quadratic equation by completing the square, one will add _____ to both sides.

Objective 1

Use the square root property to solve each equation.

5. $x^2 = 36$

5. _____

6. $(x + 7)^2 = 16$

6. _____

7. $x^2 + 4 = 9$

7. _____

8. $(x+2)^2 = 10$

8. _____

9. $(x-3)^2 = -9$

9. _____

10. $(x-4)^2 = -12$

10. _____

Objective 2

Solve each equation by completing the square.

11. $x^2 + 6x + 2 = 0$

11. _____

12. $x^2 + 8x - 6 = 0$

12. _____

13. $x^2 + 3x + 5 = 0$

13. _____

14. $3y^2 + 6y - 4 = 0$

14. _____

15. $x^2 + 6x + 12 = 0$

15. _____

16. $x^2 + 4x + 8 = 0$

16. _____

17. $2x^2 + 6x + 11 = 0$

17. _____

Objective 3

Use the formula $A = P(1 + r)^t$ to solve.

 18. Find the rate r at which \$3000 compounded annually grows to \$4320 in 2 years.

18. _____

19. Find the rate r at which \$4200 compounded annually grows to \$8232 in 2 years.

19. _____

Concept Extension

20. An isosceles triangle has legs of equal length. If the hypotenuse is 48 centimeters long, find the length of each leg.

20. _____

21. The area of a square is 36 square centimeters. Find the length of the diagonal of the square.

21. _____

Section 11.2 Solving Quadratic Equations by the Quadratic Formula

Learning Objectives
1. Solve quadratic equations by using the quadratic formula.
2. Determine the number and type of solutions of a quadratic equation by using the discriminant.
3. Solve geometric problems modeled by quadratic equations.

Objective 1

Use the quadratic formula to solve each equation.

1. $x^2 + 16x + 15 = 0$

1. _____

2. $x^2 + 14x + 45 = 0$

2. _____

3. $2x^2 + 3x - 6 = 0$

3. _____

4. $3x^2 + 12x - 20 = 0$

4. _____

5. $-2x^2 + x - 5 = 0$

5. _____

6. $\dfrac{2}{5}y^2 + \dfrac{1}{5}y + \dfrac{3}{5} = 0$

6. _____

Use the discriminant to determine the number and types of solutions of each equation.

7. $2x^2 + 4x - 5 = 0$

7. _____

8. $x^2 + 3x + 7 = 0$

8. _____

9. $6 = 4x - 5x^2$

9. _____

10. $15x^2 + 13x - 27 = 0$

10. _____

Objective 3

Solve.

 11. Nancy, Thelma, and John Varner live on a corner lot. Often, neighborhood children across their lot to save walking distance. Give the diagram below, approximate to the nearest foot how many feet of walking distance is saved from cutting across their property instead of walking around the lot.

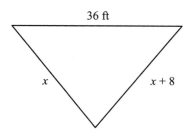

11. _____

12. A rectangle is 6 more meters long as it is high. If the area of the rectangle is 22 square meters. Find dimensions of the rectangle.

12. _____

Concept Extension

Use the quadratic formula to solve each quadratic equation.

13. $x^2 + \sqrt{5}x - 3 = 0$

13. _____

14. $3x^2 - \sqrt{6}x + 2 = 0$

14. _____

Section 11.3 Solving Equations by Using Quadratic Methods

Learning Objectives
1. Solve various equations that are quadratic in form.
2. Solve problems that lead to quadratic equations.

Objective 1

Solve.

1. $x = \sqrt{6x - 5}$

1. _____

2. $\sqrt{17 - x} = (x + 3)$

2. _____

3. $2x - 3 = \sqrt{3x}$

3. _____

4. $\dfrac{2}{x} + \dfrac{3}{x-1} = 1$

4. _____

5. $\dfrac{2}{x^2} = \dfrac{4}{x+1}$

5. _____

6. $\dfrac{4}{x+3} - \dfrac{2}{x-3} = 2$

6. _____

7. $\dfrac{5}{x^2 + 7x + 12} = \dfrac{3}{x+4} - \dfrac{6}{x+3}$

7. _____

8. $x^4 - 64 = 0$

8. _____

9. $x^4 + 3x^2 - 4 = 0$

9. _____

10. $x^{2/3} - 3x^{1/3} - 10 = 0$

10. _____

11. $x + 2x^{1/2} - 5 = 0$

11. _____

12. $(x-1)^2 - 3(x-1) - 6 = 0$

12. _____

Objective 2

Solve.

13. A jogger ran 3 miles, decreased her speed by 1 mile per hour and then ran another 4 miles. If her total time jogging was $1\frac{3}{5}$ hours, find her speed for each part of her run.

13. _____

14. The product of a number and 11 more than that number is 80. Find the number.

14. _____

15. Joe and Sarah can complete a job if they work together in 8 hours. If it takes Sarah 2 more hours than Joe to do it alone, find how long it will take each of them to complete the job separately to the nearest tenth of an hour.

15. _____

Concept Extension

Solve.

16. $x^3 - 125 = 0$

16. _____

17. $x^{-2} + 2x^{-1} + 5 = 0$

17. _____

Section 11.4 Nonlinear Inequalities in One Variable

Learning Objectives
1. Solve polynomial inequalities of degree 2 or greater.
2. Solve inequalities that contain rational expressions with variables in the denominator.

Objective 1

Solve each quadratic inequality. Write the solution in interval notation.

1. $(x-3)(x+4) \le 0$

1. _____

2. $(x-5)(x+1) > 0$

2. _____

3. $x^2 + 4x - 12 < 0$

3. _____

4. $(x+4)(x-3)(x+1) \ge 0$

4. _____

5. $\left(x^2 - 25\right)\left(x^2 - 1\right) > 0$

5. _____

6. $(2x - 3)(3x + 5)(x - 4) \le 0$

6. _____

Objective 2

Solve each inequality. Write the solution set in interval notation.

7. $\dfrac{2}{x + 3} < 6$

7. _____

8. $\dfrac{x + 3}{x - 4} \ge 0$

8. _____

9. $\dfrac{x^2}{14-5x} < 1$ 9. _____

10. $\dfrac{x(x-3)}{(x+2)(x-5)} \geq 0$ 10. _____

Concept Extension

11. Explain why $(x+5)(x-4) > 0$ and $\dfrac{x+5}{x-4} > 0$ have the same solution.

       11. _____

12. Twice a number add to 3 times its reciprocal will result in a positive solution.

       12. _____

Section 11.5 Quadratic Functions and Their Graphs

Learning Objectives
1. Graph quadratic functions of the form $f(x) = x^2 + k$.

2. Graph quadratic functions of the form $f(x) = (x-h)^2$.

3. Graph quadratic functions of the form $f(x) = (x-h)^2 + k$.

4. Graph quadratic functions of the form $f(x) = ax^2$.

5. Graph quadratic functions of the form $f(x) = a(h-k)^2 + k$

Objective 1

Sketch the graph of each quadratic function. Label the vertex, and sketch and label the axis of symmetry.

1. $f(x) = x^2 + 2$

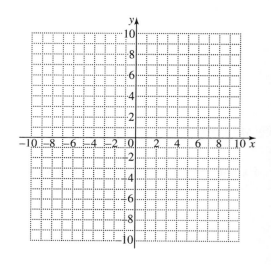

2. $f(x) = x^2 - 4$

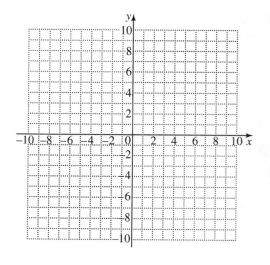

Martin-Gay *Beginning and Intermediate Algebra, Fourth Edition*

3. $f(x) = x^2 + 6$

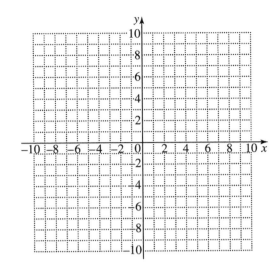

Objective 2

Sketch the graph of each quadratic equation. Label the vertex, and sketch and label the axis of symmetry.

4. $h(x) = (x+2)^2$

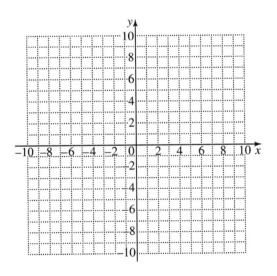

5. $f(x) = (x-4)^2$

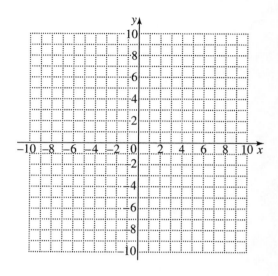

6. $f(x) = (x+3)^2$

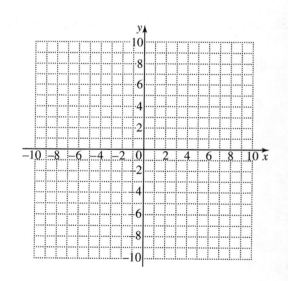

Objective 3

Sketch the graph of each quadratic equation. Label the vertex, and sketch and label the axis of symmetry.

7. $f(x) = (x-2)^2 + 5$

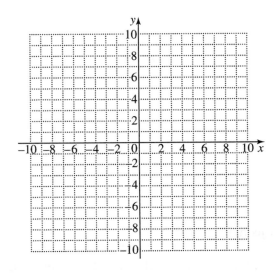

8. $f(x) = (x+1)^2 - 4$

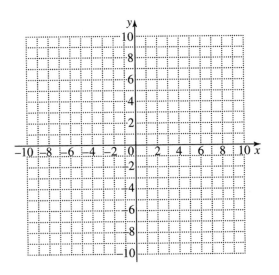

9. $f(x) = (x-3)^2 - 2$

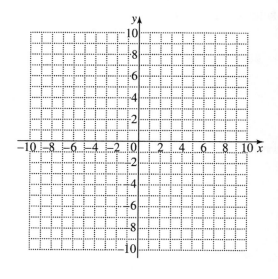

Objective 4

Sketch the graph of each quadratic equation. Label the vertex, and sketch and label the axis of symmetry.

10. $f(x) = 3x^2$

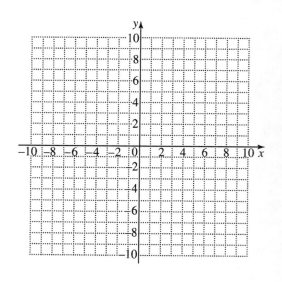

11. $f(x) = -4x^2$

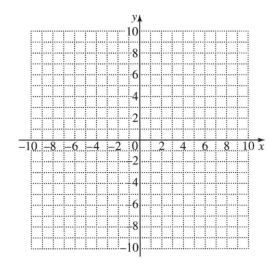

12. $f(x) = -\dfrac{1}{2}x^2$

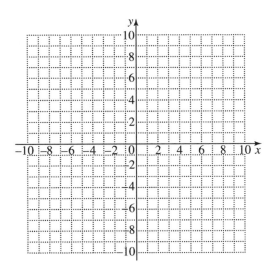

13. $f(x) = \dfrac{1}{4}x^2$

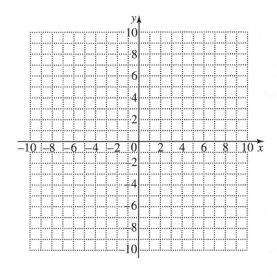

Objective 5

Sketch the graph of each quadratic equation. Label the vertex, and sketch and label the axis of symmetry.

 14. $h(x) = -3(x+3)^2 + 1$

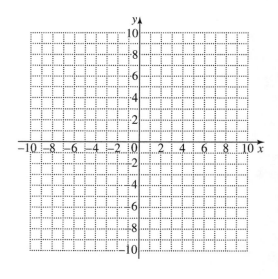

15. $f(x) = 2(x-1)^2 + 2$

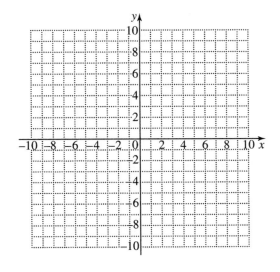

16. $f(x) = \dfrac{1}{2}(x+2)^2 + 3$

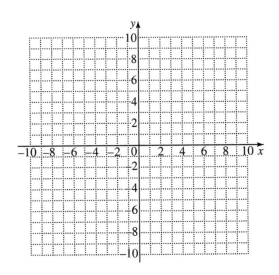

Concept Extension

17. Write the equation of the parabola that has the same shape of $f(x) = -2x^2$ but with the vertex of $(4, -6)$.

17. _____

18. What about $(-7, 9)$?

18. _____

Section 11.6 Further Graphing of Quadratic Functions

Learning Objectives
 1. Write quadratic functions in the form $y = a(x-h)^2 + k$.
 2. Derive the formula for finding the vertex of a parabola.
 3. Find the minimum or maximum value of a quadratic function.

Objective 1

Rewrite the following functions in the form $y = a(x-h)^2 + k$. State the vertex of each function.

1. $f(x) = x^2 - 2x - 15$ 1. _____

2. $f(x) = x^2 + 6x + 17$ 2. _____

3. $f(x) = x^2 + 5x + 18$ 3. _____

4. $f(x) = 2x^2 + 6x - 11$ 4. _____

5. $f(x) = -3x^2 + 12x + 8$ 5. _____

6. $f(x) = -x^2 + 8x - 9$ 6. _____

Objective 2

7. Take the equation $y = ax^2 + bx + c$ and using the variables a, b, and c to rewrite the equation to be in the form $y = a(x - h)^2 + k$. What value did you get in place of the h?

7. _____

8. The value you created from # 7 is the x-value of your vertex, how would you get the y-value of the vertex?

8. _____

Using the formula you found in # 7 and # 8, find the vertex of the following functions.

9. $y = x^2 + 4x + 7$

9. _____

10. $f(x) = 2x^2 + 6x + 9$

10. _____

11. $f(x) = 3x^2 - 6x - 13$

11. _____

12. $f(x) = -x^2 + 4x - 5$

12. _____

Objective 3

Solve.

13. If a projectile is fired straight upward from the ground with an initial speed of 96 feet per second, then its height h in feet after t seconds is given by the equation $h(t) = -16t^2 + 96t$. Find the maximum height of the projectile.

13. _____

14. The local high school bowling team is selling candles to raise money for their scholarship fund. The profit in dollars is $P(x) = 420x - x^2$.

   a. How many candles must be sold to maximize profit.

                                           14a. _____

   b. Find the maximum profit.

                                           14b. _____

Concept Extension

Find the maximum or minimum value of each function. Approximate to two decimal points. State whether it is a maximum or minimum value.

15. $f(x) = 1.2x^2 + 3.6x + 4.5$ 15. _____

16. $f(x) = -0.3x^2 + 1.6x - 1.2$ 16. _____

Chapter 11 Vocabulary

abulary Word	Definition	Example
re root property	If $a^2 = b$ then $a = \pm\sqrt{b}$	$\sqrt{36} = \pm 6$
dratic Formula	$x = \dfrac{-b \pm \sqrt{b^2 - 4ac}}{2a}$	
dratic function	$y = a(x-h)^2 + k$ Vertex: (h,k) if a is positive parabola opens upward if a is negative parabola opens downward.	$y = 4(x-3)^2 + 5$ Vertex: $(3,5)$, parabola opens upward since $a = 4$
dratic equation	$y = ax^2 + bx + c$ Vertex $\left(-\dfrac{b}{2a}, f\left(-\dfrac{b}{2a}\right)\right)$	$y = 2x^2 - 8x + 9$ Vertex at $(2,1)$

Practice Test A

Solve each equation for the variable.

1. $x^2 - 3x - 5 = 0$ 1. _____

2. $35x^2 + 46x - 16 = 0$ 2. _____

3. $x^2 + 6x = -3$ 3. _____

4. $(y-7)^2 = 36$ 4. _____

5. $a^4 = 8a^2 + 9$ 5. _____

6. $(x-2)^2 - 16(x-2) + 63 = 0$ 6. _____

7. $x^3 + 2x^2 = 9x + 18$ 7. _____

8. $\dfrac{x}{x+4} - \dfrac{2}{x-4} = \dfrac{3}{x^2-16}$ 8. _____

Solve the following functions by completing the square.

9. $x^2 - 12x + 19 = 0$ 9. _____

10. $4c^2 - 24c + 15 = 0$ 10. _____

Solve each inequality for x. Write the solution set in interval notation.

11. $(3x+5)(x-8) \geq 0$

11. _____

12. $\dfrac{(x+7)}{(2x-9)} < 0$

12. _____

13. $2x^2 - 3 > -x$

13. _____

Graph each function. Label the vertex.

14. $f(x) = (x+3)^2 + 6$

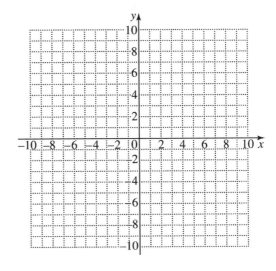

15. $h(x) = -5x^2 - 4$

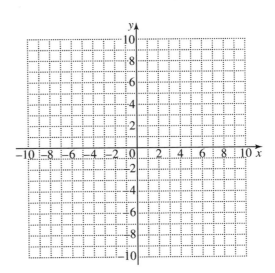

16. $g(x) = x^2 + 4x + 9$

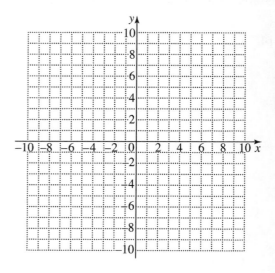

17. $f(x) = -2x^2 + 16x - 25$

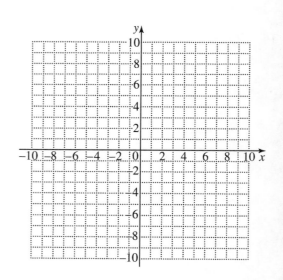

Solve.

18. One number is eight more than another number, if the product of these two number is 273, what are the two numbers.

18. _____

19. A triangle has a height of $x-8$ and a base of $2x-8$. Given that the area of the triangle is 32 feet, find the base and the height of the triangle.

19. _____

20. The height an object follows when it is shot out of a cannon follows the path of a parabola. The equation for height is given to be $h(t) = -16t^2 + 144t$, where t is time in second and h is height in feet. At what time will the object be at its maximum height?

20. _____

Practice Test B

Solve each equation for the variable.

1. $3x^2 + 20 = 19x$

    a. $4/3$ and 5   b. 4 and 5

    c. 20 and $1/3$ d. $20/3$ and 1

1. _____

2. $(x+7)^2 = 18$

    a. -1 and - 4   b. 11 and -25

    c. $7 \pm 2\sqrt{3}$ d. $-7 \pm 3\sqrt{2}$

2. _____

3. $r^2 - 3r - 54 = 0$

    a. -9 and - 6   b. 9 and -6

    c. 9 and 6   d. -9 and 6

3. _____

4. $x^2 + 3x + 7 = 0$

    a. $\dfrac{-3 \pm \sqrt{19}i}{2}$ b. $\dfrac{-3 \pm \sqrt{37}}{2}$

    c. 7 and 1   d. $\dfrac{-2 \pm \sqrt{33}i}{6}$

4. _____

5. $\dfrac{x}{x-3} - \dfrac{5}{x-4} = \dfrac{-3}{x^2 - 7x + 12}$

    a. 6   b. 6 and 3

    c. 3   d. $\varnothing$

5. _____

6. $6(x+7)^2 - 5(x+7) - 4 = 0$ 6. _____

    a. $-\dfrac{1}{2}$ and $\dfrac{4}{3}$ b. -7 and -4

    c. $-\dfrac{15}{2}$ and $-\dfrac{17}{3}$ d. 7 and 4

7. $w^4 + 8w^2 + 12 = 0$ 7. _____

    a. $\pm\sqrt{6}$ and $\pm\sqrt{2}$ b. 6 and 2

    c. 4 and 3 d. $\pm\sqrt{6}i$ and $\pm\sqrt{2}i$

8. $2x^3 + 3x^2 - 8x - 12 = 0$ 8. _____

    a. $6, -1$, and 2 b. ±2 and $-\dfrac{3}{2}$

    c. $3, 4$, and -1 d. $4, 3$, and -2

Solve the following functions by completing the square.

9. $x^2 + 10x - 13 = 0$ 9. _____

    a. $-5 \pm 2\sqrt{3}$ b. -3 and 13

    c. $-5 \pm \sqrt{38}$ d. -13 and 1

10. $-3c^2 + 36c - 18 = 0$ 10. _____

    a. $-6 \pm \sqrt{30}$ b. -6 and 3

    c. -4 and 14   d. 6 and 1

Solve each inequality for x. Write the solution set in interval notation.

11. $(x-2)(4x+5) < 0$ 11. _____

    a. $\left(-\frac{5}{4}, 2\right)$ b. $\left(-\infty, -\frac{5}{4}\right) \cup (2, \infty)$

    c. $\left[-2, \frac{5}{4}\right]$ d. $\left(-\infty, -2\right] \cup \left[\frac{5}{4}, \infty\right)$

12. $\dfrac{(6x+4)}{(x-3)} \geq 0$ 12. _____

    a. $\left[-\frac{2}{3}, 3\right)$ b. $\left(-\infty - 3\right) \cup \left(\frac{2}{3}, \infty\right)$

    c. $\left(-\infty, -\frac{2}{3}\right] \cup (3, \infty)$ d. $\left(-3, \frac{2}{3}\right)$

13. $x^2 + 5x < 6$ 13. _____

    a. $(-1, \infty)$ b. $(-\infty - 1) \cup (6, \infty)$

    c. $(-\infty, 6)$ d. $(-1, 6)$

Find the vertex of the following quadratic equations.

14. $f(x) = 5(x-4)^2 - 22$

14. _____

    a. $(5,22)$ b. $(-20,-22)$

    c. $(4,-22)$ d. $(4,22)$

15. $g(x) = 16x^2 + 120$

15. _____

    a. $(0,-120)$ b. $(0,120)$

    c. $(-16,120)$ d. $(16,120)$

Find the axis of symmetry of the following quadratic equation.

16. $h(x) = x^2 - 6x - 12$

16. _____

    a. $x = -12$ b. $x = -6$

    c. $x = -3$ d. $x = 3$

Find the y-intercept of the following quadratic equation.

17. $g(x) = -2(x-5)^2 - 11$

17. _____

    a. $(0,-61)$ b. $(0,-5)$

    c. $(0,-2)$ d. $(0,-11)$

Solve.

18. Find two consecutive even integers whose product is 840.

18. _____

    a. 42 and 20   b. 28 and 30

    c. 21 and 40   d. 35 and 24

19. The cost function for a company to make dolls is $C(x) = 4x^2 - 240x + 12000$, where x is the number of dolls sold. Find the number of dolls that need to be made to minimize costs.

19. _____

    a. 45 dolls   b. 40 dolls

    c. 30 dolls   d. 20 dolls

20. The quotient of five and eight more than four times a number, yields the quotient of the number and the sum of the number and two. Find the number.

20. _____

    a. -2 or $\frac{5}{4}$ b. -2 or 5

    c. 2 or 5   d. -5 or $\frac{1}{2}$

Chapter 12 Exponential and Logarithmic Functions
Section 12.1 The Algebra of Functions; Composite Functions

Learning Objectives
  1. Add, subtract, multiply, and divide functions.
  2. Construct composite functions.

Objective 1

For the functions f and g, find a. $(f+g)(x)$, b. $(f-g)(x)$, c. $(f \cdot g)(x)$, and d. $\left(\dfrac{f}{g}\right)(x)$.

1. $f(x) = x - 5$, $g(x) = 2x + 3$

1a. _____

1b. _____

1c. _____

1d. _____

2. $f(x) = x^2 + 1$, $g(x) = 5x$

2a. _____

2b. _____

2c. _____

2d. _____

3. $f(x)=\sqrt{x+5}, \; g(x)=4x^2$

3a. _____

3b. _____

3c. _____

3d. _____

4. $f(x)=8x+3, \; g(x)=\sqrt[3]{x}$

4a. _____

4b. _____

4c. _____

4d. _____

Objective 2

If $f(x)=x^2+3x-4$, $g(x)=2x-5$ and $h(x)=\sqrt[3]{x}$ find the following compositions.

5. $(f \circ g)(4)$

5. _____

6. $(g \circ h)(-64)$

6. _____

7. $(g \circ f)(-6)$

7. _____

Find $(f \circ g)(x)$ and $(g \circ f)(x)$.

8. $f(x) = x^2 + 1,\ g(x) = 5x$

8. _____

9. $f(x) = x - 7,\ g(x) = \sqrt{x + 3}$

9. _____

10. $f(x) = 3x^2 - 4x + 5,\ g(x) = \sqrt{x}$

10. _____

11. $f(x) = x^2 - 9,\ g(x) = \sqrt[3]{2x + 5}$

11. _____

If $f(x) = 3x,\ g(x) = \sqrt{x}$, and $h(x) = x^2 + 2$, write each function as a composition using two of the given functions.

12. $H(x) = \sqrt{x^2 + 2}$

12. _____

Concept Extension

Find the following given these values:

$$f(-1)=6 \qquad g(-1)=3$$
$$f(2)=9 \qquad g(2)=-1$$
$$f(5)=10 \qquad g(5)=-4$$
$$f(6)=13 \qquad g(6)=-6$$

13. $(f+g)(5)$

13. _____

14. $(f-g)(2)$

14. _____

15. $(f \cdot g)(-1)$

15. _____

16. $\left(\dfrac{f}{g}\right)(6)$

16. _____

17. $(g \circ f)(-1)$

17. _____

Section 12.2 Inverse Functions

Learning Objectives
1. Determine whether a function is a one-to-one function.
2. Use the horizontal line test to decide whether a function is a one-to-one function.
3. Find the inverse of a function.
4. Find the equation of the inverse of a function.
5. Graph functions and their inverses.
6. Determine whether two functions are inverses of each other.

Vocabulary
Use the choices to complete each statement.

Horizontal   **Identical**   **Inverse of** f
Reversed   **Vertical** x

$y = x$ $\dfrac{1}{f}$

1. The graphs of a function and its inverse are symmetric about the_____ line.

2. The ordered pair $(5,4)$ in a function will be _____ in its inverse.

3. To tell if a function is one-to-one it must pass the _____ line test.

4. The symbol f^{-1} means _____.

5. Two functions are inverses if both $\left(f \circ f^{-1}\right)$ and $\left(f^{-1} \circ f\right)(x)$ result to be _____.

Objective 1 and 3

Determine whether each function is one-to-one to function. If it is one-to-one, list the inverse function by switching the coordinates.

6. $f = \{(11,12),(4,3),(3,4),(6,6)\}$ 6. _____

7. $g = \{(3,2),(4,6),(5,-2),(-1,6)\}$ 7. _____

Objective 2

Determine whether the graph of each function is the graph of a one-to-one function.

8.

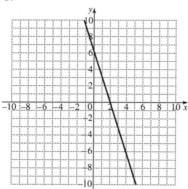

8.

9.

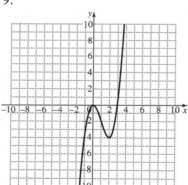

9. _____

Objective 4

Find the inverse of the following functions.

10. $f(x) = 2x - 3$

10. _____

11. $f(x) = \dfrac{x+3}{4}$

11. _____

12. $f(x) = \dfrac{5}{3x+1}$

12. _____

13. $f(x) = \sqrt[3]{x-5}$

13. _____

Objective 5

Each of the following functions is one-to-one. Find the inverse of each function and graph the function and its inverse on the same set of axes.

 14. $f(x) = 2x - 3$

14.

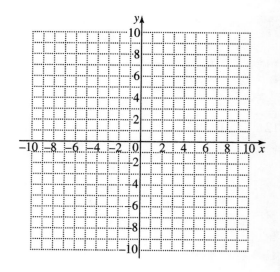

15. $f(x) = 5x + 1$

15.

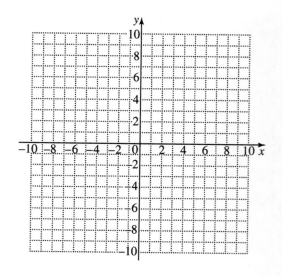

Objective 6

Solve.

16. If $f(x) = 3x + 2$, show that $f^{-1}(x) = \dfrac{1}{3x+2}$.

16. _____

Concept Extension

17. Suppose that f is a one-to-one function and that $f(-5) = 25$.

    a. Write the corresponding ordered pair.

17a. _____

    b. Name one of the ordered pair that is a solution of f^{-1}

17b. _____

Section 12.3 Exponential Functions

Learning Objectives
1. Graph exponential functions.
2. Solve equations of the form $b^x = b^y$.
3. Solve the problems modeled by exponential equations.

Objective 1

Graph each exponential function.

1. $y = 2^x$

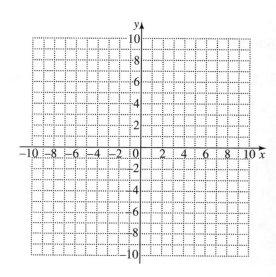

2. $y = \left(\dfrac{1}{4}\right)^x$

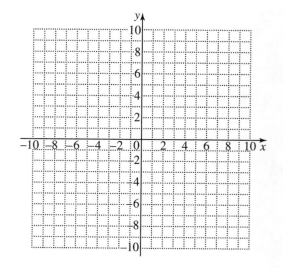

3. $y = 3^{x+2}$

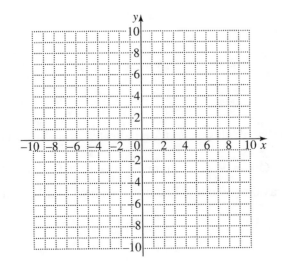

Objective 2

Solve each equation for x.

4. $2^x = 8$

4. _____

5. $5^x = \dfrac{1}{125}$

5. _____

6. $128 = 2^{6x}$

6. _____

7. $27^{x+1} = 9$

7. _____

Solve. Unless otherwise indicated, round results to one decimal place.

 8. One type of uranium has a daily radioactive decay rate of 0.4%. If 30 pounds of this uranium is available today, find how much will still remain after 50 days. Use $y = 30(2.7)^{-0.004t}$, and let t be 50.

8. _____

9. A bank account grows at 8% interest being compounded continuously. How much money will be in an account if it started with $25,000 and the money will sit in the account for 25 years. Use $A = 25000(2.7)^{.08t}$, and let t = 25.

9. _____

10. The decline of a population of rats is modeled by the equation $P = 8,000(2.7)^{-0.008t}$. How large will the population be if $t = 15$?

10. _____

Concept Extension

11. Graph the function: $y = 2^{|x|}$.

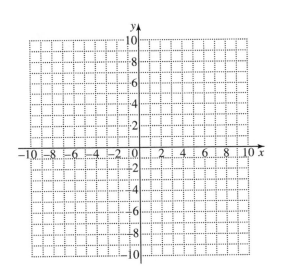

Section 12.4 Logarithmic Functions

Learning Objectives
1. Write exponential equations with logarithmic notation and write logarithmic equations with exponential notation.
2. Solve logarithmic equations by using exponential notation.
3. Identify and graph logarithmic functions.

Objective 1

Write each as an exponential equation.

1. $\log_4 16 = 2$

1. _____

2. $\log_2 \dfrac{1}{8} = -3$

2. _____

3. $\log_7 x = 9$

3. _____

4. $\log_6 \sqrt[3]{6} = \dfrac{1}{3}$

4. _____

Write each as a logarithmic equation.

5. $5^{-3} = \dfrac{1}{125}$

5. _____

6. $2^{1/4} = \sqrt[4]{2}$

6. _____

7. $3^4 = 81$

7. _____

8. $\pi^7 = z$

8. _____

Find the value of each logarithmic expression.

9. $\log_2 32$

9. _____

10. $\log_{12} 1$

10. _____

11. $\log_6 \dfrac{1}{36}$

11. _____

Objective 2

Solve.

12. $\log_x 100 = 2$

12. _____

13. $\log_3 \dfrac{1}{27} = x$

13. _____

14. $\log_2 64 = x$

14. _____

15. $\log_5 5^8 = x$

15. _____

16. $6^{\log_6 9} = x$

16. _____

Simplify.

 17. $2^{\log_2 3}$

17. _____

18. $\log_3 9$

18. _____

Objective 3

Graph each logarithmic function. Label any intercepts.

 19. $y = \log_3 x$

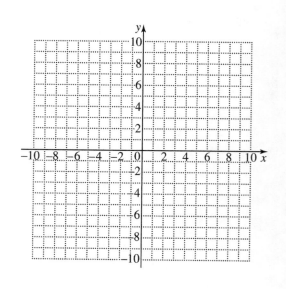

20. $f(x) = \log_5 x$

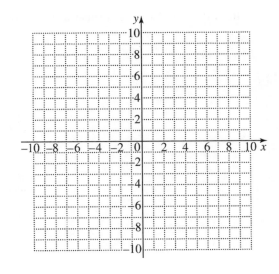

21. $f(x) = \log_{1/2} x$

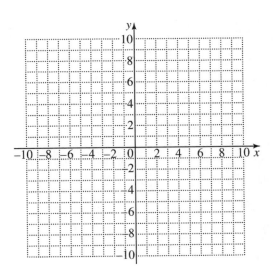

Concept Extension

Graph the following two functions on the same grid. Is there a relationship between the two of them?

22. $f(x) = \log_4 x$ $\qquad\qquad$ $g(x) = 4^x$

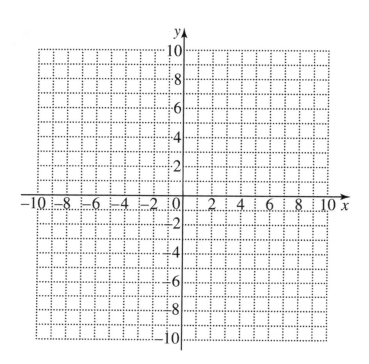

Section 12.5 Properties of Logarithms

Learning Objectives
1. Use the product property of logarithms.
2. Use the quotient property of logarithms.
3. Use the power property of logarithms.
4. Use the properties of logarithms together.

Objective 1

Write each sum as a single logarithm. Assume the variables represent positive numbers.

1. $\log_3 x + \log_3 5$ 1. _____

2. $\log_7 2x + \log_7 (x+1)$ 2. _____

3. $\log_7 (x-2) + \log_7 x^4$ 3. _____

4. $\log_{10} 5 + \log_{10} 2 + \log_{10} (x^2 + 2)$ 4. _____

Objective 2

Write each difference as a single logarithm. Assume that variables represent positive numbers.

5. $\log_3 x - \log_3 6$ 5. _____

6. $\log_{11} 5x - \log_{11}(x+6)$

6. _____

7. $\log_2(x^2 - 6) - \log_2(x+5)$

7. _____

8. $\log_9 150 - \log_9 25$

8. _____

Objective 3

Use the power property to rewrite each expression.

9. $\log_4 x^6$

9. _____

10. $\log_6 7^3$

10. _____

11. $\log_5 \sqrt{y}$

11. _____

12. $\log_8 10^{-5}$

12. _____

Objective 4

Write each as a single logarithm.

13. $\log_4 2 + \log_4 10 - \log_4 5$ 13. _____

14. $\log_2 (x+2) + 3\log_2 x - \log_2 (x-5)$ 14. _____

15. $4\log_8 x - 3\log_8 x + 2\log_8 2$ 15. _____

16. $\frac{1}{2}\log_6 x + 4\log_6 x - 2\log_6 (x+2)$ 16. _____

Write each expression as a sum or difference of logarithms. Assume that variables represent positive numbers.

17. $\log_3 5x^2$ 17. _____

18. $\log_6 \dfrac{8y}{x^3}$

18. _____

19. $\log_7 \dfrac{\sqrt{x}}{(x+4)^2}$

19. _____

20. $\log_5 \sqrt[3]{\dfrac{4x}{7}}$

20. _____

Concept Extension

21. If $\log_b 3x = 1 + \log_b x$, find b.

21. _____

Section 12.6 Common Logarithms, Natural Logarithms, and Change of Base

Learning Objectives
1. Identify common logarithms and approximate them by calculator.
2. Evaluate common logarithms of powers of 10.
3. Identify natural logarithms and approximate them by calculator.
4. Evaluate natural logarithms of powers of e.
5. Use the change of base formula.

Objective 1

Use a calculator to approximate each logarithm to four decimal places.

1. $\log 2.56$

1. _____

2. $\log 0.028$

2. _____

Objective 2

Find the exact value.

3. $\log 1000$

3. _____

4. $\log 0.0000001$

4. _____

5. $\log \dfrac{1}{10,000}$

5. _____

Solve each equation for x. Give an exact solution and a four-decimal-place approximation.

6. $\log x = 2.2$

6. _____

7. $\log 2x = 1.1$

7. _____

8. $\log(3x - 2) = -0.3$

8. _____

Objective 3

Use a calculator to approximate each logarithm to four decimal places.

  9. $\ln 5$

9. _____

10. $\ln 0.061$

10. _____

Objective 4

Find the exact value.

11. $\ln e^{0.02}$

11. _____

12. $\ln \sqrt[5]{e}$

12. _____

13. $\ln \dfrac{1}{e^5}$

13. _____

14. Find how much money Dana Jones has after 12 years if $1400 is invested at 8% interest compounded continuously. Use the formula $A = Pe^{rt}$ to solve.

14. _____

Objective 5

Approximate each logarithm to four decimal places.

15. $\log_4 9$

15. _____

16. $\log_{1/2} 12$

16. _____

17. $\log_6 0.023$

17. _____

Concept Extension

18. Without using a calculator, which of the following is larger? $\log 80$ or $\ln 80$

18. _____

Section 12.7 Exponential and Logarithmic Equations and Applications

Learning Objectives
1. Solve exponential equations.
2. Solve logarithmic equations.
3. Solve problems that can be modeled by exponential and logarithmic equations.

Objective 1

Solve each equation. Give an exact solution, and also approximate the solution to four decimal places.

1. $5^x = 9$ 1. _____

2. $2^{x-3} = 5$ 2. _____

3. $6^{3x} = 24$ 3. _____

4. $5^{2x-1} = 60$ 4. _____

Objective 2

Solve each equation.

5. $\log_3(x+5)=2$

6. $\log_4 x + \log_4(x+6)=2$

7. $\ln 8 + \ln(x+1)=0$

8. $2\log x - \log x^3 = 4$

9. $\ln 4 + \ln x = 0$

Objective 3

Solve. Round answers to the nearest whole number.

10. The size of the wolf population as Isle Royale National Park increases at a rate of 4.3% per year. If the size of the current population is 83 wolves, find how many there should be in 5 years. Use $y = y_0 e^{0.043t}$ and round to the nearest whole.

10. _____

11. Find how long it will take $750 to triple if it is invested at 8% interest compounded quarterly. Use the formula $A = P\left(1 + \dfrac{r}{n}\right)^{nt}$ to solve this compound interest problem.

11. _____

12. A bacterial culture grows according to the formula $P = P_0 a^t$. If it takes 4 days for the culture to double in size, how long will it take it to triple?

12. _____

Concept Extension

Without solving the following equations find the values of x that cannot be a solution.

13. $\log(x-5)-\log(x^2+7)=0$

13. _____

14. $\log(x^2+1)-\log(x+4)=0$

14. _____

Chapter 12 Vocabulary

Vocabulary Word	Definition	Example
The Algebra of Functions • Sum • Difference • Product • Quotient • Composite	• $(f+g)(x)=f(x)+g(x)$ • $(f-g)(x)=f(x)-g(x)$ • $(f \cdot g)(x)=f(x) \cdot g(x)$ • $\left(\dfrac{f}{g}\right)(x)=\dfrac{f(x)}{g(x)}$ • $(f \circ g)(x)=f(g(x))$	$f(x)=3x \qquad g(x)=x+2$ • $(f+g)(x)=3x+x+2=4x+2$ • $(f-g)(x)=3x-(x+2)=2x-2$ • $(f \cdot g)(x)=3x \cdot (x+2)=3x^2+6x$ • $\left(\dfrac{f}{g}\right)(x)=\dfrac{3x}{x+2}$ • $(f \circ g)(x)=3(x+2)=3x+6$
One-to-one function	A function that not only for ever x there is only one y, but for every y there is only one x. Must pass BOTH vertical and horizontal line test.	$y=3x+2$ is a one-to-one function.
Inverse function	The inverse of f is denoted as f^{-1}. Ordered pairs are reversed in the inverse function.	$y=4x+3$ and $y=\dfrac{x-3}{4}$ are inverses.
Exponential function	$f(x)=b^x$; $b>0,\ b \neq 0$	$f(x)=7^x$
Logarithmic function	$f(x)=\log_b x$; $b>0,\ b \neq 1$	$f(x)=\log_3 5$
Properties of logarithms • Product Property • Quotient Property • Power Property	• $\log_b xy = \log_b x + \log_b y$ • $\log_b \dfrac{x}{y} = \log_b x - \log_b y$ • $\log_b x^r = r \log_b x$	• $\log_2 5y = \log_2 5 + \log_2 y$ • $\log_2 \dfrac{6x}{7} = \log_2 6x - \log_2 7$ • $\log_4 x^6 = 6 \log_4 x$

Common Logarithm	$\log x = \log_{10} x$	
Natural Logarithm	$\ln x = \log_e x$	
Continuously Compound Interest Formula	$A = Pe^{rt}$; r is the annual interest rate for P dollars invested for t years	
Logarithm Property of Equality	$\log_b a = \log_b c$ then $a = c$	$\log_4 x = \log_4 5$ $x = 5$

Practice Test A

If $f(x) = 3x - 4$ and $g(x) = -2x$, find the following.

1. $(f + g)(x)$ 1. _____

2. $(f \cdot g)(x)$ 2. _____

3. $(f \circ g)(x)$ 3. _____

4. $(g \circ f)(4)$ 4. _____

Find the inverse of the following functions.

5. $f(x) = \dfrac{1}{2}x - 7$ 5. _____

6. $f(x) = x^3 - 4$

6. _____

7. $f(x) = \dfrac{5x + 8}{7}$

7. _____

Use the properties of logarithms to write each expression as a single logarithm.

8. $2\log_6 x + \log_6 3 - \log_6(x + 2)$

8. _____

9. $\dfrac{1}{2}\log_3 x - 3\log_3 2 + \log_3 5$

9. _____

10. Approximate $\log_5 12$ to the four decimal places.

10. _____

Solve each logarithmic equation for x.

11. $\log_4 x = 5$

11. _____

12. $\ln \sqrt[5]{e} = x$

12. _____

13. $\log_2 \left(x^2 - 2x \right) = 3$

13. _____

14. $\log_3 \left(x + 5 \right) - \log_3 \left(x - 2 \right) = 2$

14. _____

15. Solve $\ln \left(2x - 5 \right) = 1.68$ accurate to four decimal places.

15. _____

Graph.

16. $y = \left(\dfrac{1}{5}\right)^x$

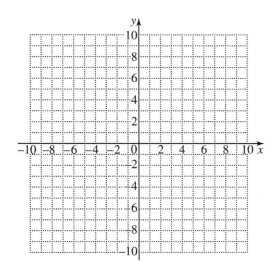

17. $y = 4^x$

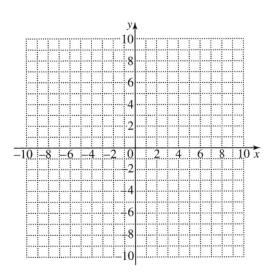

Solve.

18. Find the amount in the account if $6000 is invested at 8.5% interested compounded quarterly for 7 years. Use the formula $A = P\left(1 + \dfrac{r}{n}\right)^{nt}$.

18. _____

19. Find the amount of money that needs to be put in an account at 6% compounded continuously, if it is necessary to have $12,000 in 3 years. Use the formula $A = Pe^{rt}$

19. _____

20. A radioactive material decays according to the formula $A = A_0 (2)^{-t}$. How much of a 6-gram sample will be left in 4 years.

20. _____

Practice Test B

If $f(x)=6x$ and $g(x)=4-2x$, find the following.

1. $(f-g)(x)$ 1. _____

      a. $4x+4$ b. $8x-4$

      c. $4-8x$ d. $4x-4$

2. $(f \cdot g)(x)$ 2. _____

      a. $24x-2$ b. $4-12x^2$

      c. $24x-12x^2$ d. $24x-2x^2$

3. $(f \circ g)(x)$ 3. _____

      a. $24-12x$ b. $4-12x$

      c. $4-12x^2$ d. $24x-12x^2$

4. $(g \circ f)(-2)$ 4. _____

      a. 28 b. 32

      c. 42 d. 48

Find the inverse of the following functions.

5. $f(x) = 4x + 9$ 5. _____

   a. $\dfrac{x}{4} - 9$ b. $4x - 9$

   c. $\dfrac{x}{9} - 4$ d. $\dfrac{x-9}{4}$

6. $f(x) = \sqrt[3]{x} + 2$ 6. _____

   a. $\sqrt[3]{x} + 2$ b. $\sqrt[3]{x-2}$

   c. $x^3 - 2$ d. $\sqrt[3]{x} - 2$

7. $f(x) = \dfrac{5 - 3x}{2}$ 7. _____

   a. $\dfrac{2x+3}{-5}$ b. $\dfrac{2x-5}{-3}$

   c. $\dfrac{2x-5}{3}$ d. $\dfrac{2x+5}{3}$

Use the properties of logarithms to write each expression as a single logarithm.

8. $\log_3 6 - 2\log_3 3 + 5\log_3 x$ 8. _____

   a. $\log_3 \dfrac{10x}{3}$ b. $\log_3 \dfrac{2}{3x^5}$

   c. $\log_3 \dfrac{2x^5}{3}$ d. $\log_3 \dfrac{3x^5}{2}$

9. $\log_5 (x-1) - 6\log_5 x + \dfrac{1}{2}\log_5 2$ 9. _____

   a. $\log_5 \dfrac{\sqrt{2}(x-1)}{x^6}$ b. $\log_5 \dfrac{(x-1)}{6x}$

   c. $\log_5 \dfrac{6x}{\sqrt{2}(x-1)}$ d. $\log_5 \dfrac{x^6}{\sqrt{2}(x-1)}$

10. Approximate $\log_8 0.025$ to the four decimal places.   10. _____

   a. -1.2763 b. -1.7740

   c. 1.6742 d. 2.3321

Solve each logarithmic equation for x.

11. $\log_x 64 = 6$

11. _____

    a. -4 b. 4

    c. 2 d. -2

12. $\log x = 1.5$

12. _____

    a. 31.6228 b. -24.5672

    c. 32.3498 d. 28.7429

13. $\log_6 (3x + 8) = 2$

13. _____

    a. 11 b. $-\frac{8}{3}$

    c. -12 d. $\frac{32}{3}$

14. $\log_2 x - \log_2 (x - 2) = 4$

14. _____

    a. $\frac{8}{3}$ b. $\frac{32}{15}$

    c. $\frac{11}{5}$ d. 2

Martin-Gay *Beginning and Intermediate Algebra, Fourth Edition*

15. Solve $\ln(4x+5)=0.025$ accurate to four decimal places.

15. _____

    a. -0.9937 b. -0.9832

    c. 0.9928 d. 0.9952

Graph.

16. $y=3^x$

16. _____

a.
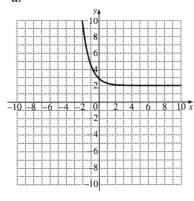

b.
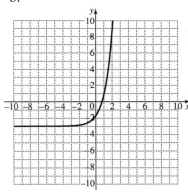

c.
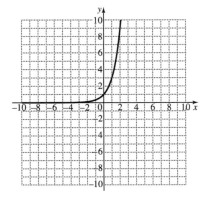

d.
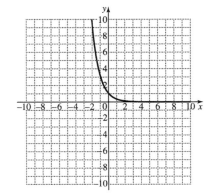

17. $y = \left(\dfrac{1}{2}\right)^x$

17. _____

a.

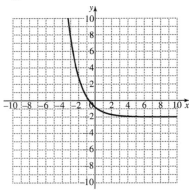

b.

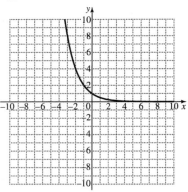

c.

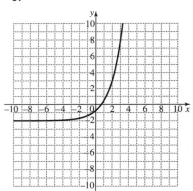

d.

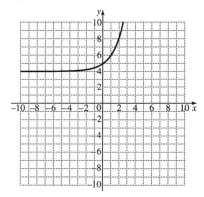

Solve.

18. Find the amount in the account if $4500 is invested at 6% interested compounded quarterly for 10 years. Use the formula $A = P\left(1 + \dfrac{r}{n}\right)^{nt}$.

18. _____

a. $8214.46

b. $8220.34

c. $8127.50

d. $8163.08

19. Find the amount of money to which a $1250 investment will grow to if it is invested at 9.5% compounded continuously for 5 years. Use the formula $A = Pe^{rt}$

19. _____

a. $1899.43

b. $1998.67

c. $2010.02

d. $2105.34

20. The spread of foot and mouth disease through a herd of cattle can be modeled using the formula $P = P_0 e^{0.25t}$, where t is the number of days. If a rancher does not act quickly to treat three cases, how many cattle will have the disease in 2 weeks?

20. _____

a. 99 cows

b. 95 cows

c. 102 cows

d. 89 cows

Chapter 13 Conic Sections
Section 13.1 The Parabola and the Circle

Learning Objectives

1. Graph parabolas of the form $x = a(y-k)^2 + h$ and $y = a(x-h)^2 + k$.
2. Graph circles of the form $(x-h)^2 + (y-k)^2 = r^2$.
3. Write the equation of a circle, given its center and radius.
4. Find the center and the radius of a circle, given its equation.

Vocabulary

Use the choices to complete each statement.

Center	**Circle**	**Conic Sections**
Diameter	**Radius**	**Vertex**

1. Half a circle's diameter is its _____.

2. For a parabola that opens downward, the highest point is called the _____.

3. The circle and parabola are examples of _____.

4. The distance from one side of the circle to the other, running through the center, is the

   _____.

5. A _____ is the set of all points equidistant for a fixed point called the

   _____ .

Objective 1

The graph of each equation is a parabola. Find the vertex of the parabola and sketch the graph.

 6. $x = (y-2)^2 + 3$

6. _____

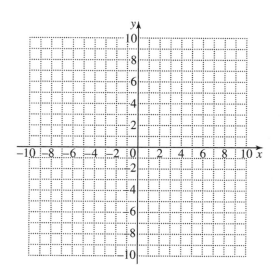

7. $y = -2(x+1)^2 - 4$

7. _____

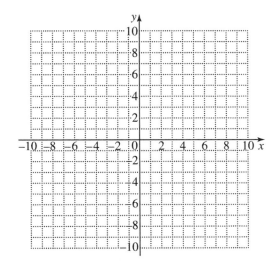

8. $x = 3y^2 + 6y + 2$

8. _____

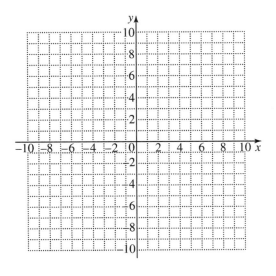

9. $y = x^2 - 6x + 3$

9.  _____

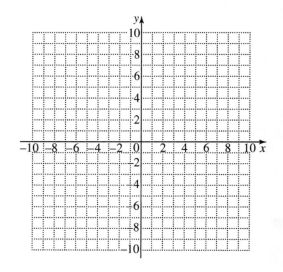

Objective 2

The graph of each equation is a circle. Find the center and the radius, and then sketch.

 10. $(x-5)^2 + (y+2)^2 = 1$

10.  _____

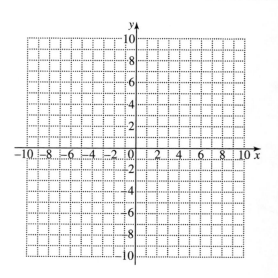

Martin-Gay *Beginning and Intermediate Algebra, Fourth Edition*

11. $x^2 + (y-3)^2 = 4$

11. _____

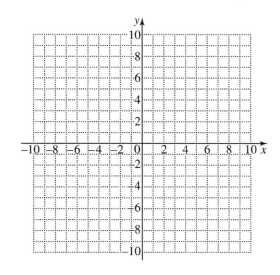

12. $x^2 + 6x + y^2 - 8y = 0$

12. _____

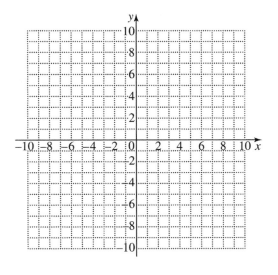

13. $x^2 + y^2 + 4x + 2y = 11$

13. _____

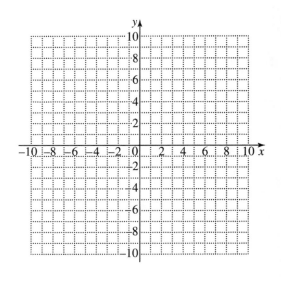

Objective 3

Write an equation of the circle with the given center and radius.

14. $(3, 4); 5$

14. _____

15. $(-1, -6); 7$

15. _____

16. $(-5, 4); 3\sqrt{5}$

16. _____

Objective 4

Find the center and the radius of the given circle equations.

17. $4x^2 + 4y^2 = 64$

17. _____

18. $x^2 + 8x + y^2 + 12y = 10$

18. _____

19. $x^2 + y^2 - 6x - 10y = 4$

19. _____

Concept Extension

20. Analyze the problems from this section, how can you tell by first glance if the equation is a parabola or a circle.

20. _____

21. Under what conditions would the graph of the $x = a(y - k)^2 + h$ have no y-intercepts.

21. _____

Section 13.2 The Ellipse and the Hyperbola

Learning Objective
1. Define and graph an ellipse.
2. Define and graph a hyperbola.

Vocabulary
Use the choices to complete each statement.

Center	Ellipse	Foci
Hyperbola	$(0,a)$ **and** $(0,-a)$	$(b,0)$ **and** $(-b,0)$

1. Two fixed points on a hyperbola and ellipse are referred to as _____.

2. A(n) _____ is a set of points such that the sum of their distances to the foci is a constant.

3. The graph $\dfrac{x^2}{a^2} - \dfrac{y^2}{b^2} = 1$ is a(n) _____ with the center at the origin and x-intercepts at

   _____.

4. The _____ is the midpoint between the two foci.

Objective 1

Sketch the graph of each equation.

5. $x^2 + \dfrac{y^2}{4} = 1$

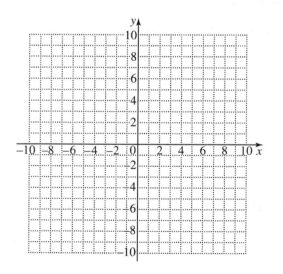

6. $\dfrac{x^2}{16} + \dfrac{y^2}{25} = 1$

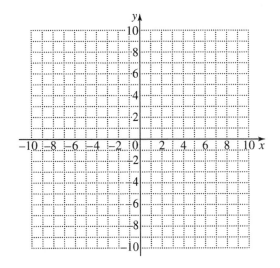

7. $16x^2 + y^2 = 64$

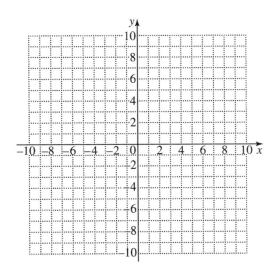

8. $\dfrac{(x-1)^2}{4} + \dfrac{(y-1)^2}{25} = 1$

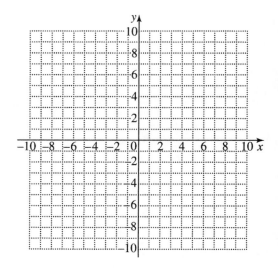

9. $\dfrac{(x+4)^2}{36} + \dfrac{(y+2)^2}{49} = 1$

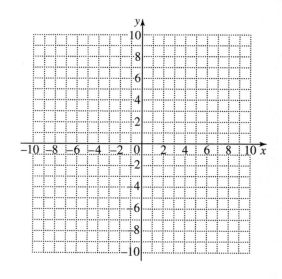

Objective 2

Sketch the graph of each equation.

10. $\dfrac{x^2}{4} - y^2 = 1$

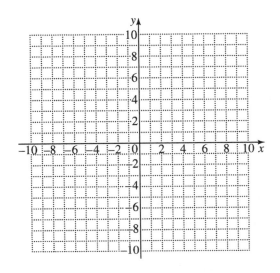

11. $\dfrac{x^2}{25} - \dfrac{y^2}{16} = 1$

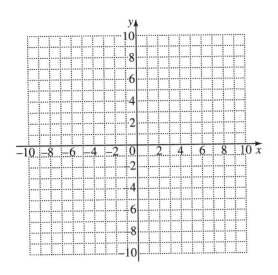

12. $9y^2 - x^2 = 36$

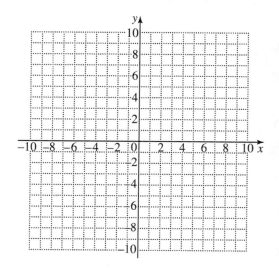

13. $\dfrac{(x-2)^2}{25} - \dfrac{(y-1)^2}{16} = 1$

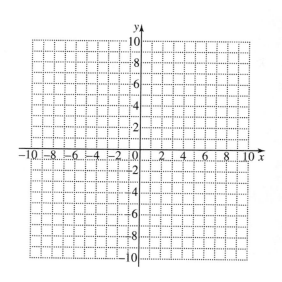

Martin-Gay *Beginning and Intermediate Algebra, Fourth Edition*

4. $\dfrac{(y+1)^2}{4} - \dfrac{(x+2)^2}{9} = 1$

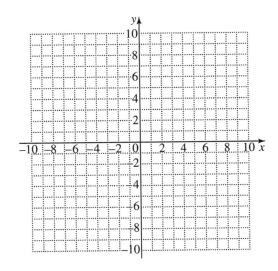

Concept Extension

Graph the following equations.

15. $4x^2 + 9y^2 - 16x - 18y = 11$

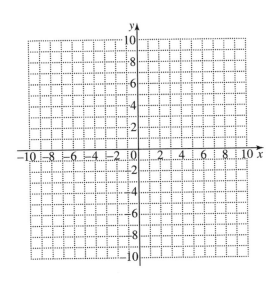

16. $x^2 + 4y - y^2 - 2x = 12$

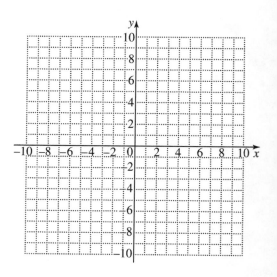

Section 13.3 Solving Nonlinear Systems of Equations

Learning Objectives
1. Solve a nonlinear system by substitution.
2. Solve a nonlinear system by elimination.

Objective 1

Solve each nonlinear system of equations for real solutions using substitution.

1. $\begin{cases} y^2 = 4 - x \\ x - 2y = 4 \end{cases}$

1. _____

2. $\begin{cases} x^2 + y^2 = 16 \\ 2x + 3y = 0 \end{cases}$

2. _____

3. $\begin{cases} 3x + y = 4 \\ xy = 2 \end{cases}$

3. _____

4. $\begin{cases} y = x^2 + 25 \\ y = 2x^2 + 9 \end{cases}$

4. _____

5. $\begin{cases} x^2 + 4y^2 = 16 \\ x - y = 4 \end{cases}$

5. _____

Objective 2

Solve each nonlinear system of equations for real solutions using elimination.

 6. $\begin{cases} 2x^2 + 3y^2 = 14 \\ -x^2 + y^2 = 3 \end{cases}$

6. _____

7. $\begin{cases} y = x^2 + 12 \\ y = -x^2 + 10 \end{cases}$

7. _____

8. $\begin{cases} x^2 + y^2 = 36 \\ y = x^2 - 16 \end{cases}$

8. _____

9. $\begin{cases} x^2 + 4y^2 = 16 \\ x^2 + (y-1)^2 = 9 \end{cases}$

9. _____

10. $\begin{cases} x + y = 4 \\ y = x^2 + 4 \end{cases}$

10. _____

Concept Extension

Solve.

11. The sum of the squares of two numbers is 36. The sum of the first number and twice the second number is 6. Find the two numbers.

11. _____

12. How many possible real solutions are there for the intersection of a circle and an ellipse?

12. _____

Section 13.4 Nonlinear Inequalities and Systems of Inequalities

Learning Objectives
1. Graph a nonlinear inequality.
2. Graph a system of nonlinear inequalities.

Objective 1

Graph each inequality.

1. $\dfrac{x^2}{4} + \dfrac{y^2}{9} \leq 1$

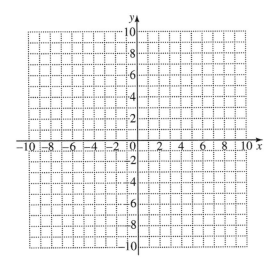

2. $\dfrac{x^2}{16} - y^2 > 1$

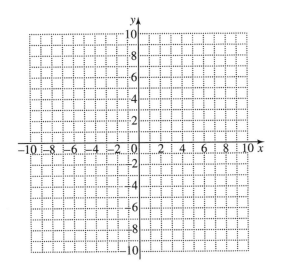

3. $x^2 + (y-2)^2 < 36$

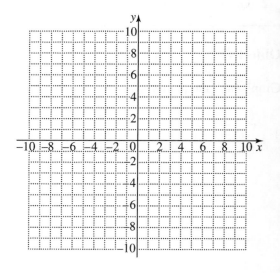

4. $y > x^2 + 2x$

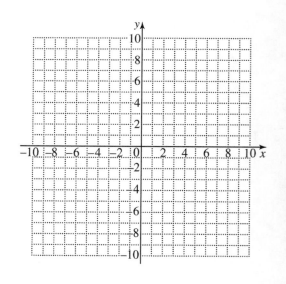

Martin-Gay *Beginning and Intermediate Algebra, Fourth Edition*

Objective 2

Graph each system.

5. $\begin{cases} x^2 + y^2 > 9 \\ y > x^2 \end{cases}$

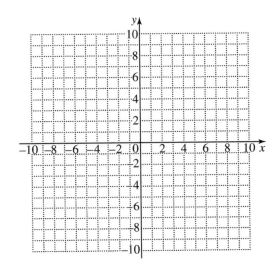

6. $\begin{cases} x^2 - y^2 \geq 1 \\ x + y < 2 \end{cases}$

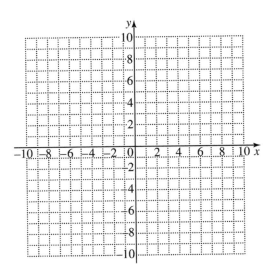

7. $\begin{cases} \dfrac{x^2}{16} + \dfrac{y^2}{25} > 1 \\ y < x^2 \end{cases}$

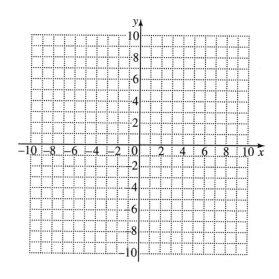

8. $\begin{cases} x^2 + y^2 \le 25 \\ x^2 + y^2 > 4 \end{cases}$

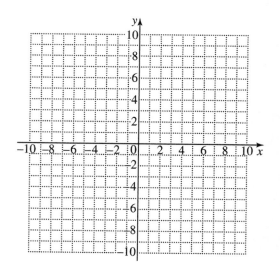

Concept Extension

Graph the following system.

9.

$$\begin{cases} y \geq x^2 \\ y < 2x + 1 \\ x \geq 3 \\ y < 4 \end{cases}$$

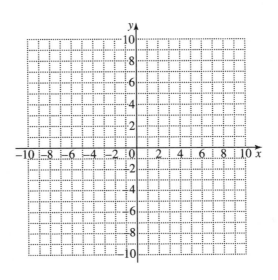

Chapter 13 Vocabulary

Vocabulary Word	Definition	Example
Parabolas	$y = a(x-h)^2 + k$ $x = a(y-k)^2 + h$	
Circle	$(x-h)^2 + (y-k)^2 = r^2$	
Ellipse	$\dfrac{x^2}{a^2} + \dfrac{y^2}{b^2} = 1$	
Hyperbola	$\dfrac{x^2}{a^2} - \dfrac{y^2}{b^2} = 1$ or $\dfrac{x^2}{b^2} - \dfrac{y^2}{a^2} = 1$	
Nonlinear System of Equations	A system of at least one that is not linear.	

Practice Test A

Sketch the graph of each equation.

1. $x^2 + y^2 = 64$

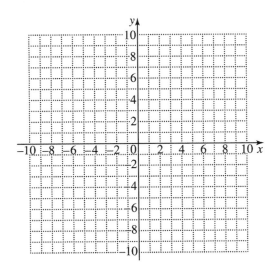

2. $x^2 - \dfrac{y^2}{16} = 1$

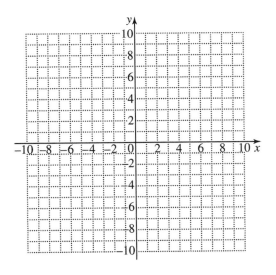

3. $y = (x - 4)^2 + 6$

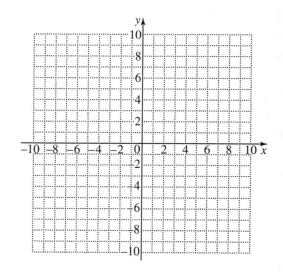

4. $\dfrac{(x+1)^2}{25} + \dfrac{(y-2)^2}{4} = 1$

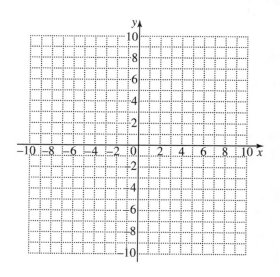

5. $x^2 + 8x - y^2 - 4y = 5$

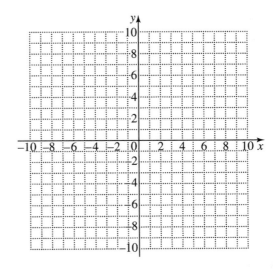

Solve each system.

6. $\begin{cases} x^2 - y^2 = 16 \\ 2x^2 + y^2 = 5 \end{cases}$

6. _____

7. $\begin{cases} 4x^2 + 4y^2 = 16 \\ x = y + 2 \end{cases}$

7. _____

8. $\begin{cases} y^2 - x^2 = 4 \\ x = 3y \end{cases}$

8. _____

Graph the solution of each system.

9. $\begin{cases} 3x + 4y > 12 \\ y < x^2 + 1 \end{cases}$

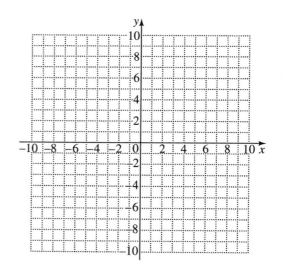

10. $\begin{cases} \dfrac{x^2}{16} - y^2 < 1 \\ x + y \geq 3 \end{cases}$

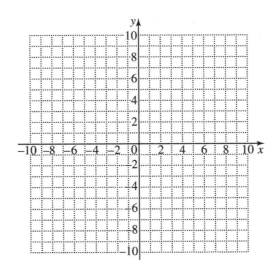

11. $\begin{cases} x^2 + y^2 < 9 \\ x^2 + y^2 > 1 \end{cases}$

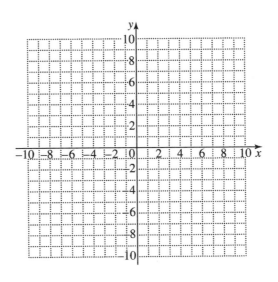

12. The cross section of the satellite antenna is a parabola given by the equation $y = \dfrac{1}{36}x^2$, with distances measured in feet. If the dish is 20 feet wide, how deep is it? Round your answer to one decimal point.

12. _____

Practice Test B

Sketch the graph of each equation.

1. $x^2 + y^2 = 25$

1. _____

a.

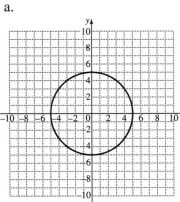

b.

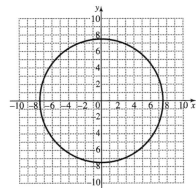

c.

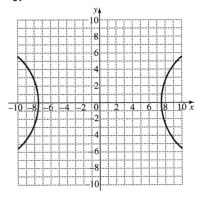

d.

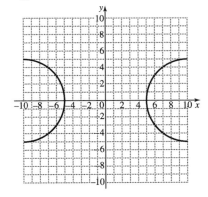

2. $\dfrac{y^2}{9} - x^2 = 1$

2. _____

a.

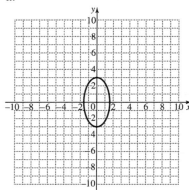

b.

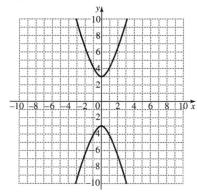

c.

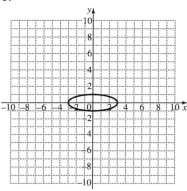

d.

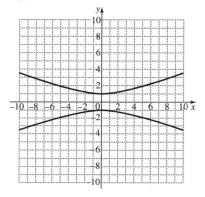

3. $x = (y-1)^2 - 3$ 3. _____

a.

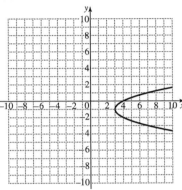

b.

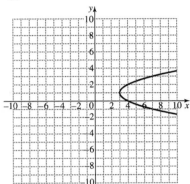

c.

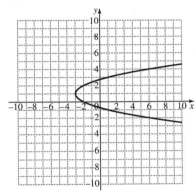

d.

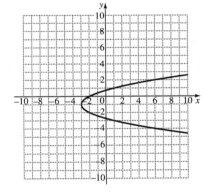

4. $\dfrac{(x-1)^2}{49} - \dfrac{(y+4)^2}{16} = 1$

4. _____

a.

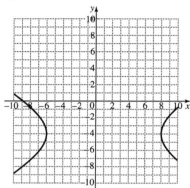

b.

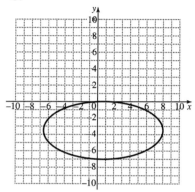

c.

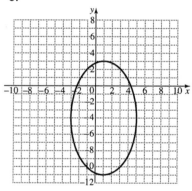

d.

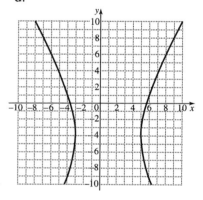

Martin-Gay *Beginning and Intermediate Algebra, Fourth Edition*

5. $x^2 + 16x + y^2 - 10x = 12$ 5. _____

a.

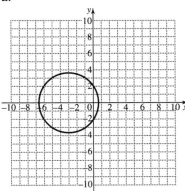

b.

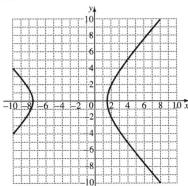

c.

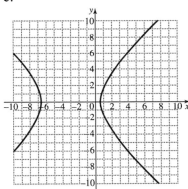

d.

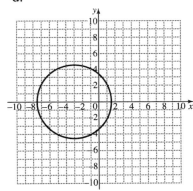

Solve each system.

6. $\begin{cases} x^2 + y^2 = 12 \\ -x^2 + y^2 = 8 \end{cases}$ 6. _____

a. $(2, -10)(2, 10)$

b. $\left(-\sqrt{2}, \pm\sqrt{10} \right)\left(\sqrt{2}, \pm\sqrt{10} \right)$

c. $\left(2, \pm 2\sqrt{2} \right)\left(-2, \pm 2\sqrt{2} \right)$

d. $\varnothing$

7. $\begin{cases} \dfrac{x^2}{4} - y^2 = 1 \\ 2x - y = 4 \end{cases}$

7. _____

    a. $\left(4, \pm\sqrt{3}\right)$ b. $\left(2,0\right)\left(4,\sqrt{3}\right)$

    c. $\left(\dfrac{34}{15}, \dfrac{8}{15}\right)\left(2,0\right)$ d. $\varnothing$

8. $\begin{cases} 2x^2 = y - 3 \\ x = -2y \end{cases}$

8. _____

    a. $\left(1,5\right)\left(-2,1\right)$ b. $\left(-2,1\right)\left(2,1\right)$

    c. $\left(4,-2\right)\left(2,-1\right)$ d. $\varnothing$

Graph the solution of each system.

9. $\begin{cases} x > y^2 - 4 \\ y < x^2 - 1 \end{cases}$

9. _____

a.

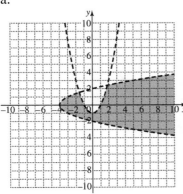

b.

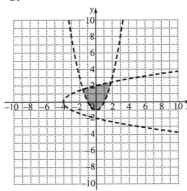

c.

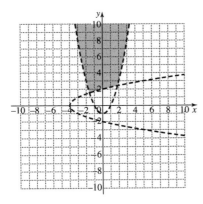

d.

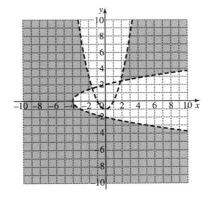

10. $\begin{cases} \dfrac{x^2}{4} + \dfrac{y^2}{9} < 1 \\ x - 2y \geq 6 \end{cases}$

10. _____

a.

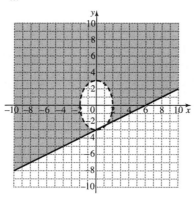

b.

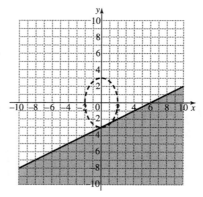

c.

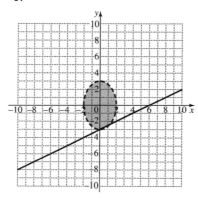

d.

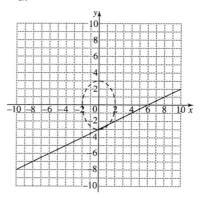

11. $\begin{cases} 4x^2 + y^2 < 64 \\ x^2 + y^2 \geq 1 \end{cases}$

11. _____

a.

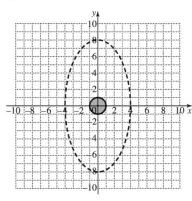

b.

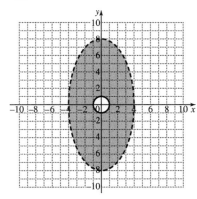

c.

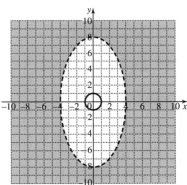

d.

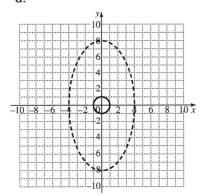

12. The area of a ellipse $\dfrac{x^2}{a^2} + \dfrac{y^2}{b^2} = 1$ is given by the equation $A = \pi ab$. Find the area of the ellipse $16x^2 + 9y^2 = 144$.

12. _____

a. 144π b. 12π

c. 288π d. 25π

Chapter 14 Sequences, Series, and the Binomial Theorem
Section 14.1 Sequences

Learning Objectives
  1. Write the terms of a sequence given the general term.
  2. Find the general term of a sequence.
  3. Solve applications that involve sequences.

Vocabulary
Use the choices to complete each statement.
Finite   **General Term**   **Infinite**

1. A(n) _____ sequence is a function whose domain is $\{1,2,3,...\}$.

2. The _____, or a_n, is the nth term of the sequence.

3. A(n) _____ sequence is a function whose domain is $\{1,2,3,...,n\}$

Objective 1

Write the first five terms of each sequence whose general term is given.

4. $a_n = n + 2$ 4. _____

5. $a_n = (-3)^{n-1}$ 5. _____

6. $a_n = n^2 - 4$ 6. _____

7. $a_n = \dfrac{1}{n+3}$

7. _____

8. $a_n = (-1)^n (2n)$

8. _____

Find the indicated term for each sequence whose general term is given.

9. $a_n = -2n; \ a_9$

9. _____

10. $a_n = \dfrac{n-4}{n+2}; \ a_{10}$

10. _____

11. $a_n = \dfrac{(-1)^{n+1}}{n^2}; \ a_{20}$

11. _____

12. $a_n = (n-3)(n+4)$; a_{34} 12. _____

13. $a_n = 2^{3n-1}$; $a_n = 8$ 13. _____

Objective 2

Find the general term a_n for each sequence whose first four terms are given.

14. $3, -9, 27, -81$ 14. _____

 15. $-2, -4, -8, -16$ 15. _____

16. $\dfrac{1}{3}, \dfrac{1}{5}, \dfrac{1}{7}, \dfrac{1}{9}$

16. _____

17. $\dfrac{3}{5}, \dfrac{2}{3}, \dfrac{5}{7}, \dfrac{3}{4}$

17. _____

Objective 3

Solve.

18. Mrs. Laser agrees to give her son Mark an allowance of \$0.10 on the first day of his 14-day vacation, \$0.20 on the second day, \$0.40 on the third day, and so on. Write an equation of a sequence whose terms correspond to Mark's allowance. Find the allowance Mark will receive on the last day of his vacation.

18. _____

19. George puts \$80 in his savings account. Each paycheck, he will put an additional \$40 more in the box. Find the amount in his account for the first five paychecks after the deposits.

19. _____

Concept Extension

Find the first five terms of each sequence. Round each term after the first to four decimal places.

20. $a_n = \dfrac{\sqrt{n+2}}{\sqrt{n}+2}$

20. _____

21. $a_n = \left(1 + \dfrac{0.025}{n}\right)^{n-1}$ Round to four decimal places of needed.

21. _____

Section 14.2 Arithmetic and Geometric Sequences

Learning Objectives
1. Identify arithmetic sequences and their common differences.
2. Identify geometric sequences and their common ratios.

Vocabulary

Use the choices to complete each statement.

Arithmetic   **Difference**   **First**
Geometric   **Last**   **Ratio**

1. A(n) _____ sequence is one in which each term (after the first) differs from the

   term before it by a constant common _____.

2. A(n) _____ sequence is one in which each term (after the first) is derived from

   multiplying the proceeding term by a constant common _____.

3. In a sequence the term a_1 is the _____ term.

Objective 1

Write the first five terms of the arithmetic or geometric sequence whose first term a_1, and common difference, d are given.

4. $a_1 = 6; \ d = -2$ 4. _____

5. $a_1 = 25; \ d = 7$ 5. _____

6. $a_1 = -2; \ d = -5$ 6. _____

7. $a_1 = 10;\ d = 16$ 7. _____

Find the indicated term of each sequence.

8. The ninth term of the arithmetic sequence whose first term is 13 and whose common difference is 9.

8. _____

9. The eleventh term of the arithmetic sequence whose first term is 4 and whose common difference is 3.

9. _____

10. The sixth term of the arithmetic sequence whose first term is 1 and whose common difference is – 6.

10. _____

 11. The twenty-fifth term of the arithmetic sequence 20, 18, 16,…

11. _____

12. The tenth term of the arithmetic sequence 12, 8, 4, …

12. _____

Objective 2

Write the first five terms of the geometric sequence whose first term a_1, and common ratio, r are given.

13. $a_1 = 48$; $r = \dfrac{1}{2}$ 13. _____

14. $a_1 = 10$; $r = -3$ 14. _____

15. $a_1 = -1$; $r = 4$ 15. _____

Find the indicated term of each sequence.

16. The fifth term of the geometric sequence whose first term is 8 and whose common ratio is .

                                                                                        16. _____

17. The seventh term of the geometric sequence whose first terms in – 2 and whose common ratio is 3.

                                                                                        17. _____

18. The sixth term of the geometric sequence $3, 12, 48...$

18. _____

19. The eighth term of the geometric sequence $4, -2, 1..$

19. _____

Concept Extension

Write the first four terms of the following sequences.

20. $a_1 = 2560.25; \quad d = -1450.75$

20. _____

21. $a_1 = 32.3; \quad r = -2.2$

21. _____

Section 14.3 Series

Learning Objective
1. Identify finite and infinite series and use summation notation.
2. Find partial sums.

Objective 1

Evaluate.

1. $\displaystyle\sum_{i=4}^{7}(2i+4)$ 1. _____

2. $\displaystyle\sum_{i=3}^{8}4i$ 2. _____

3. $\displaystyle\sum_{i=6}^{10}i(i-2)$ 3. _____

4. $\displaystyle\sum_{i=2}^{8} \frac{2i}{i+4}$

4. _____

5. $\displaystyle\sum_{i=2}^{6} 2^{i+2}$

5. _____

Write each series with summation notation.

6. $4 + 9 + 16 + 25 + 36 + 49$

6. _____

7. $5 + 7 + 9 + 11 + 13$

7. _____

8. $6 + 11 + 18 + 27 + 38$

8. _____

Objective 2

Find each partial sum.

9. Find the sum of the first two terms of the sequence whose general term is $a_n = n(n-6)$.

9. _____

10. Find the sum of the first four terms of the sequence whose general term is $a_n = (-1)^{2n+1}$.

10. _____

11. Find the sum of the fist six terms of the sequence whose general term is $a_n = n(n-3)$.

11. _____

12. Find the sum of the first five terms of the sequence whose general term is $a_n = -\dfrac{n}{2}$.

12. _____

Concept Extension

13. Show that $\displaystyle\sum_{i=1}^{5} \frac{i^2}{i} \neq \frac{\displaystyle\sum_{i=1}^{5} i^2}{\displaystyle\sum_{i=1}^{5} i}$. What do you think $\displaystyle\sum_{i=1}^{5} \frac{i^2}{i}$ might equal? Show that it does.

13. _____

Section 14.4 Partial Sums of Arithmetic and Geometric Sequences

Learning Objectives
1. Find the partial sum of an arithmetic sequence.
2. Find the partial sum of a geometric sequence.
3. Find the sum of the terms of an infinite geometric sequence.

Objective 1

Use the partial sum formula to find the partial sum of the given arithmetic and geometric sequence.

1. Find the sum of the first six terms of the arithmetic sequence $3, 6, 9, \ldots$

1. _____

2. Find the sum of the first five positive even integers.

2. _____

3. Find the sum of the first seven terms of $-15, -8, -1 \ldots$

3. _____

4. Find the sum of the first four positive odd integers.

4. _____

Objective 2

5. Find the sum of the first five terms of the geometric sequence $4, 12, 36, \ldots$

5. _____

6. Find the sum of the first six terms of the geometric sequence $32, -16, 8, \ldots$

6. _____

7. Find the sum of the first four terms of the geometric sequence $1, \dfrac{1}{4}, \dfrac{1}{16}, \ldots$

7. _____

8. Find the sum of the first five terms of the geometric sequence if $a_1 = 16$ and $r = -\dfrac{2}{3}$.

8. _____

Objective 3

Find the sum of the terms of each infinite geometric sequence.

9. $32, 16, 8\ldots$

9. _____

10. $-10, -5, -\dfrac{5}{2}$

10. _____

11. $4, -\dfrac{1}{2}, \dfrac{1}{16}, \ldots$

11. _____

12. $3, \dfrac{1}{3}, \dfrac{1}{27}, \ldots$

12, _____

Concept Extension

13. Write $0.44\overline{4}$ as an infinite geometric series and use the formula for S_{∞} to write it as a rational number.

13. _____

14. Write $0.36\overline{36}$ as an infinite geometric series and use the formula for S_{∞} to write it as a rational number.

14. _____

Section 14.5 The Binomial Theorem

Learning Objectives
1. Use Pascal's triangle to expand binomials.
2. Evaluate factorials.
3. Use the binomial theorem to expand binomials.
4. Find the nth term in the expansion of a binomial raised to a positive power.

Objective 1

Use Pascal's triangle to expand the binomial.

1. $(m+n)^3$ 1. _____

2. $(a-b)^4$ 2. _____

3. $(x+y)^8$ 3. _____

Objective 2

Evaluate each expression.

4. $\dfrac{8!}{4!}$ 4. _____

5. $\dfrac{24!}{20!}$

5. _____

6. $\dfrac{16!}{7!9!}$

6. _____

Objective 3

Use the binomial formula to expand each binomial.

7. $(x-y)^5$

7. _____

8. $(2x+3)^6$

8. _____

9. $(5a - 2b)^4$

9. _____

10. $(4 + 3x)^6$

10. _____

Objective 4

Find the indicated term.

11. The fifth term of the expansion of $(2x + 5z)^8$.

11. _____

12. The fourth term of the expansion of $(2r - s)^5$.

12. _____

13. The third term of the expansion of $(m-3n)^7$.

13. _____

Concept Extension

14. Expand the expression $\left(\sqrt{a}-\sqrt{6}\right)^4$.

14. _____

15. Find the term containing x^3 in the expansion of $\left(\sqrt{x}+\sqrt{2}\right)^{10}$.

15. _____

Chapter 14 Vocabulary

Vocabulary Word	Definition	Example
Infinite Series	A function whose domain is the set of natural numbers $\{1,2,3,4...\}$	$1,3,5,7,....$
Finite Series	A function whose domain is the set of natural numbers $\{1,2,3,...,n\}$	$2,4,6,8,10$
Arithmetic Sequence	A sequence in which each term differ from the previous term by a common difference.	$3,6,9,12,15,...$
General term of an Arithmetic Sequence.	$a_n = a_1 + (n-1)d$	$a_n = 5 + (n-1)6$
Geometric Sequence	A sequence in which each term (after the first term) is found by multiplying the previous term by a common ratio	$8,-16,32,-64,...$
General Term of a Geometric Sequence.	$a_n = a_1 r^{n-1}$	$a_n = 3(4)^{n-1}$
Series	Sum of terms in a sequence.	$1+3+5+7$
Summation Notation	$\sum$: sigma	$\displaystyle\sum_{i=2}^{6} i = 2+3+4+5+6$
Partial Sum	S_n	S_3 of $1,2,3$ $1+2+3=6$

Partial Sum of Arithmetic Sequence	$S_n = \dfrac{n}{2}(a_1 + a_n)$	$3, 5, 7, 9....$ $S_3 = \dfrac{3}{2}(3+7) = 15$
Partial Sum of a Geometric Sequence	$S_n = \dfrac{a_1\left(1 - r^n\right)}{1 - r}$	$2, 4, 8, 16...$ $S_3 = \dfrac{2\left(1 - 2^3\right)}{\left(1 - 2\right)} = 14$
Sum of an Infinite Geometric Sequence	$S_\infty = \dfrac{a_1}{1 - r}$	$4, 2, 1, ...$ $S_\infty = \dfrac{4}{1 - \frac{1}{2}} = 8$

Practice Test A

Find the indicated term(s) of the given sequences.

1. The first six terms of the sequence $a_n = \dfrac{(2n+1)}{n+4}$.

1. _____

2. The ninth term of the sequence $a_n = 5\left(\dfrac{1}{3}\right)^{n-1}$.

2. _____

3. The general term of the sequence $7, -1, \dfrac{1}{7}, \ldots$

3. _____

4. The general term of the sequence $24, 10, -4, \ldots$

4. _____

Find the partial sum of the given sequence.

5. S_7 of the sequence $a_n = 5 + (n-1)6$

5. _____

6. S_6 of the sequence $a_n = -4\left(-\dfrac{1}{2}\right)^{n-1}$

6. _____

7. S_∞ of the sequence $a_1 = 144$ and $r = \dfrac{1}{6}$

7. _____

8. S_4 of the sequence $8, -16, 32, \ldots$

8. _____

Martin-Gay *Beginning and Intermediate Algebra, Fourth Edition*

9. $\displaystyle\sum_{i=2}^{7} i^2 - 3i + 5$

9. _____

10. $\displaystyle\sum_{i=1}^{6} 6(3)^i (-1)^i$

10. _____

Expand each binomial.

11. $(2x - y)^7$

11. _____

12. $(a + 3b)^5$

12. _____

Practice Test B

Find the indicated term(s) of the given sequences.

1. The first six terms of the sequence $a_n = \dfrac{n^2 - 2}{n + 6}$.

1. _____

   a. $7, 8, 9, 10, 11, 12$ b. $1, 2, 7, 14, 23, 34$

   c. $-\dfrac{5}{7}, \dfrac{3}{4}, \dfrac{11}{9}, \dfrac{9}{5}, \dfrac{27}{11}, \dfrac{19}{6}$ d. $-\dfrac{1}{7}, \dfrac{1}{4}, \dfrac{7}{9}, \dfrac{7}{5}, \dfrac{23}{11}, \dfrac{17}{6}$

2. The ninth term of the sequence $a_n = 16\left(\dfrac{1}{2}\right)^{n-1}$.

2. _____

   a. $\dfrac{1}{16}$ b. $\dfrac{1}{32}$

   c. 8192 d. 4096

3. The general term of the sequence $18, -9, \dfrac{9}{2}, \dots$

3. _____

   a. $a_n = 18\left(\dfrac{1}{2}\right)^{n}$ b. $a_n = 18\left(-\dfrac{1}{2}\right)^{n-1}$

   c. $a_n = -18\left(\dfrac{1}{2}\right)^{n-1}$ d. $a_n = 18 - 9(n-1)$

4. The general term of the sequence $15, -5, -25, \ldots$

4. _____

    a. $a_n = 15 + (n-1)(-20)$ b. $a_n = 15\left(\frac{1}{3}\right)^n$

    c. $a_n = 15 + (n-1)20$ d. $a_n = 15\left(-\frac{1}{3}\right)^{n-1}$

Find the partial sum of the given sequence.

5. S_7 of the sequence $a_n = 22 + (n-1)(-8)$

5. _____

    a. -12 b. -22

    c. -14 d. -18

6. S_6 of the sequence $a_n = 100\left(\frac{1}{5}\right)^{n-1}$

6. _____

    a. 613 b. $\dfrac{15624}{25}$

    c. $\dfrac{16130}{26}$ d.. 625

7. S_∞ of the sequence $a_1 = 150$ and $r = \dfrac{1}{5}$

7. _____

      a. 184   b. 191

      c. $\dfrac{401}{2}$ d. $\dfrac{375}{2}$

8. S_∞ of the sequence $2, 8, 32, \ldots$

8. _____

      a. 673   b. 682

      c. 660   d. 713

9. $\displaystyle\sum_{i=4}^{9} 3i + 6$

9. _____

      a. 156   b. 160

      c. 153   d. 147

10. $\displaystyle\sum_{i=3}^{9} i(2i+3)$ 10. _____

    a. 686   b. 724

    c. 622   d. 580

11. Find the third term of the expansion of $(4x+y)^5$.   11. _____

    a. $160x^2y^3$ b. $1280x^4y$

    c. $20xy^4$ d. $640x^3y^2$

12. Find the sixth term of the expansion of $(c-2d)^8$.   12. _____

    a. $1120c^4d^4$ b. $-1792c^3d^5$

    c. c^6 d. $1792c^2d^6$

ANSWERS

Chapter 1

Section 1.2

1. Zero

3. Irrational numbers

5. Inequality symbols

7. Rational numbers

9. Integers

11. >

13. >

15. F

17. T

19. $-8 > -15$

21. $5 \geq 4$

23. integer, real, rational

25. rational real

27. -6 pounds

29. -15 degrees

31. 5

33. 8.9

35. =

37. >

39. T

41. T

43. July

Section 1.3

1. Equivalent
3. Simplified
5. Reciprocals
7. $2 \cdot 2 \cdot 2 \cdot 2$
9. $2 \cdot 2 \cdot 3 \cdot 5$

11. $\frac{1}{2}$

13. $\frac{3}{7}$

15. $\frac{8}{7}$ or $1\frac{1}{7}$

17. 1

19. $\frac{1}{4}$

21. $\frac{7}{24}$

23. $\frac{9x}{30x}$

25. $3\frac{1}{11}$

27. No

Section 1.4

1. Subtract
3. Multiply
5. Solution
7. Base; equivalent
9. 27
11. 0.0000000032
13. 40

15. $\frac{41}{50}$

17. $\frac{5}{6}$

19. $\frac{43}{8}$ or $5\frac{3}{8}$

21. No

23. $\frac{2}{3}x + 6$

25. $3(x+8)$
27. $5 - x \geq -7$
29. Answers may vary.

Section 1.5

1. Opposites
3. –n
5. 968
7. – 165
9. – 55
11. 2.7
13. 12° F
15. – 8
17. – 17
19. – 5
21. A. x plus negative y
   B. The opposite of x plus negative y.
   C. the opposite of x minus y.

Section 1.6

1. Supplementary
3. Complementary
5. x – 23
7. x – 23
9. 19
11. $-\dfrac{41}{35}$ or $-1\dfrac{6}{35}$
13. – 27
15. – 26
17. 0
19. 9
21. 113
23. – 7
25. 124°
27. A. T
   B. F
   C. F
   D. F

Section 1.7

1. Positive
3. Reciprocal
5. Negative
7. 0
9. $-\dfrac{8}{27}$
11. - 120

13. 2
15. – 32
17. – 64
19. $-\dfrac{6}{5}$ or $-1\dfrac{1}{5}$
21. – 6
23. $\dfrac{20}{27}$
25. – 44
27. 54
29. 1

Section 1.8

1. Commutative property of addition
3. Additive inverses
5. Associative property of addition
7. Identity element of multiplication
9. 16 + x
11. 3k + 5j
13. 10(cs)
15. 4 + (9+b)
17. $\left(-\dfrac{8}{13}\cdot\dfrac{39}{16}\right)\cdot s$
19. –r + 3 + 7p
21. -23x + 70
23. 11(x+y)
25. $\dfrac{1}{2}\left(\dfrac{1}{2}t+5\right)$
27. Answers may vary.

Practice Test A

1. 2x + 8 < 15
2. $21-\dfrac{1}{2}x$
3. 162
4. $\dfrac{68}{77}$
5. – 24
6. $16\dfrac{1}{4}$ or $\dfrac{65}{4}$
7. 0
8. $\dfrac{10}{9}$ or $1\dfrac{1}{9}$
9. – 11
10. – 83

11. -8
12. $<$
13. $<$
14. $=$
15. A. $1, \sqrt{25}$

    B. $1, 0, \sqrt{25}$

    C. $12, -4, 1, 0, \sqrt{25}$

    D. $12, -4, 1, 0, \sqrt{25}, 11, 75, \frac{5}{9}$

    E. Π

    F. All of them
16. 63
17. Identity Element of Addition
18. Associative property of addition
19. Identity element of multiplication
20. Distributive property
21. $-\frac{1}{15}$
22. $\frac{5}{7}$
23. -28 pts.
24. The first contractor
25. Yes

Practice Test B
1. C
2. A
3. B
4. D
5. C
6. A
7. B
8. B
9. C
10. B
11. B
12. C
13. B
14. B
15. A
16. B
17. C
18. A
19. C

20. A
21. B
22. B
23. B
24. B
25. B

Chapter 2

Section 2.1

1. Exponent
3. Numerical coefficient
5.

$3x^4$	3	x	4
$\frac{1}{5}x^3$	$\frac{1}{5}$	x	3
$-x$	-1	x	1
7	7		0

7. $-12y - 6s$
9. $-7.5x + 6.2xy$
11. $-3y + 2x - 4z$
13. $4.3y + 5.2$
15. $8z - 27y$
17. $-3x + 18$
19. It would double.

Section 2.2

1. True
3. Addition
5. Multiplication
7. -0.7
9. -4
11. -3.75
13. $\frac{9}{10}$
15. $20 - P$
17. $2x + 4$
19. $6x + 16y$

Section 2.3

1. $-\frac{36}{17}$
3. -11

5. 9

7. $\dfrac{3}{4}$

9. $-\dfrac{18}{7}$

11. 8

13. 0

15. No solution

17. 0

19. 10

Section 2.4

1. 4

3. -17

5. -9

7. Chicago = 46
   New England = 10

9. 30 ft.

11. 31 m, 18 m

13. $14\dfrac{3}{5}$ in, $16\dfrac{3}{5}$ in, $43\dfrac{3}{5}$ in

15. $58°, 60°, 62°$

17. 8, 10, 12

19. 4, 6, 8

Section 2.5

1. 32

3. 4

5. $25°$

7. $43\dfrac{1}{3}$ mph

9. approx. 12 years

11. 78 ft by 36 ft

13. $\dfrac{2A}{h} = b$

15. $\dfrac{S - 2\pi r^2}{2\pi r} = h$

17. $\dfrac{3V}{\pi r^2} = h$

19. Answers may vary

Section 2.6

1. 60

3. 90

5. 55%

7. $78

9. 12.5%

11. $46.58

13. 11.11%

15. $53,333.33

17. 12 lbs

19. 150 L

21. Yes.

Section 2.7

1. 6.5 hrs

3. 17 hours

5. 82.5 miles

7. 15 dimes 60 nickels

9. 41 DVD players
   123 MP3 players

11. 3 times

13. $1220 at 5%
    $2580 at 7%

15. $211.11 at 10%
    $5888.89 at 8%

17. $4800 at 9%
    $3600 at 12%

19. 8.57 L

Section 2.8

1. False

3. Linear Inequality in One Variable.

5. $x \le -1$

7. $(-\infty, -1]$

9. $[6, \infty)$

11. $[-5, \infty)$

13. $(-\infty, 7)$

15. $(-7, \infty)$

17. $[-6, 15)$

19. $(-8, 46)$

21. $x > -2$

23. $x < 150$

Practice Test A

1. $-4t + 18$

2. $x - 30$

3. $-3.32p - 6.6$

4. $\dfrac{17}{28}x - \dfrac{18}{7}$

5. -8

6. -20

7. $24/5$

8. -11

9. No Solution

10. -11

11. 15

12. 430

13. All real numbers

14. 33

15. 20

16. 9, 11, 13

17. 61.67 kg.

18. 6.5 hrs.

19. 160 ft.

20. $\dfrac{S-P}{\text{Pr}} = t$

21. $m = \dfrac{y-b}{x}$

22. $\left(4\dfrac{1}{2}, \infty\right)$

23. $[4, \infty)$

24. $[6, 12]$

25. between 2 and 5 ft.

Practice Test B

1. C
2. D
3. A
4. B
5. A
6. B
7. D
8. A
9. C
10. B
11. D
12. C
13. C
14. C
15. B
16. C
17. C
18. B
19. A
20. A
21. D
22. C
23. B

24. B
25. D

Chapter 3
Section 3.1

1. y-axis, x-axis
3. Origin, (0, 0)
5. x-coordinate, y-coordinate
7. 1997 to 1998
9. Yes. Answers vary
11. 250
13.

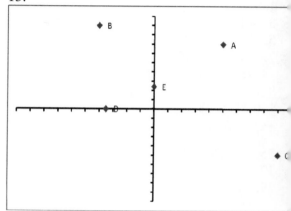

15. yes, no, no

17. (-4, -2) (4, 0)
19.

0	0
2	-8
-3	12

21. a. (-1.6, -2.2)
    b. 24.8
    c. 36.48

Section 3.2

1. yes
3. yes

5.

0	0
5	0
2	3

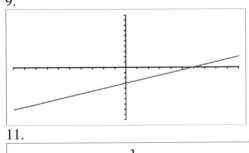

7.

0	-2
1	-5
2	-8

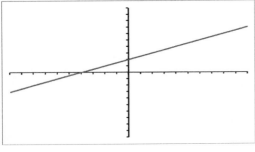

9.

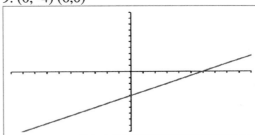

11.

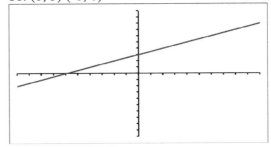

13.

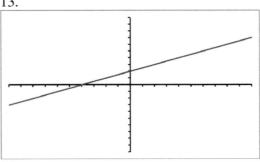

15.

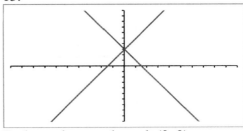

Both graphs pass through (0, 3)
17. $700

Section 3.3

1. Horizontal
3. Standard
5. Vertical
7. (0, 6) (-6, 0)
9. (0, -4) (6,0)

11. (0, 3) (-6, 0)

13. (0,0)

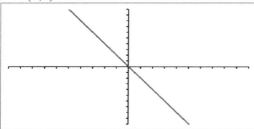

15.

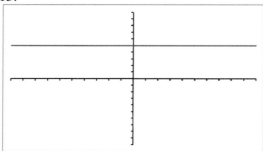

17. Graph is the y-axis.

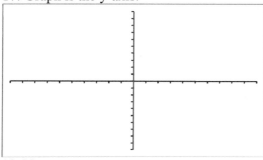

19. No. answers vary

Section 3.4

1. m, b
3. Undefined
5. Parallel
7. m = -1
9. m = undefined
11. $m = \frac{2}{3}$
13. $m = \frac{4}{5}$
15. m = 0
17. perpindicular
19. $\frac{1}{2}$
21. a. (0, 31) (22, 54)

　b. $y = \frac{14}{11}x + 31$

　c. ≈ 70.15%

　d. 2031

Section 3.5

1. Point slope
3. Slope-intercept
5. Vertical
7. $y = \frac{2}{3}x - 2$
9.

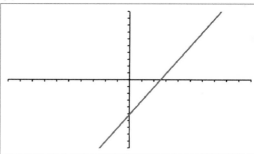

11.

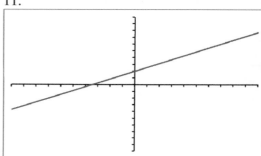

13. 8x + y = -13
15. 3x + y = 1
17. $\frac{7}{4} + y = -5$
19. x = 24
21. x = -7
23. (-8, -2); $y = \frac{2}{3}x + \frac{10}{3}$

Section 3.6

1. Vertical
3. Function
5. Range
7. {2, 8, -7, 1}
9. yes
11. yes
13. no
15. yes
17. yes
19. (-1, 10) (2, 16) (0,6)
21. Domain $(-\infty, \infty)$ Range $(-\infty, \infty)$

Martin-Gay *Beginning and Intermediate Algebra, Fourth Edition*

23. $\left(-\infty,\dfrac{3}{2}\right)\cup\left(\dfrac{3}{2},\infty\right)$

25a. 166.38 cm

  b. 148.25 cm

Practice Test A

1. (-3, 13) (0, 4) (7, -17)
2. III
3. no
4. no
5.

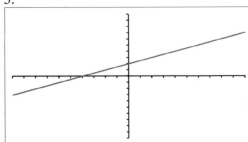

6.

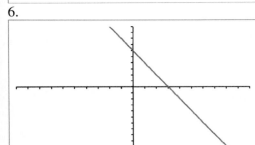

7. (0, -5) (-3, 0)
8. (0, 3) (-2, 0)

9. (0, 0)

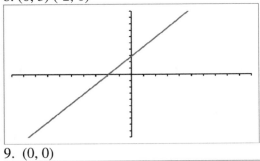

10.

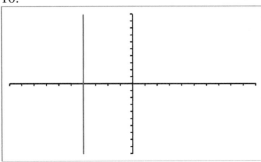

11. $m=-\dfrac{3}{2}$

12. $m=-\dfrac{1}{3}$

13. $m=\dfrac{3}{7}$

14. neither

15. $m=-\dfrac{5}{9}$ $b=-\dfrac{4}{3}$

16. $y=\dfrac{4}{7}x-6$

17.

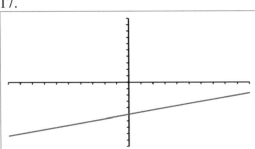

18.

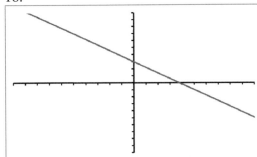

19. $4x+y=-27$
20. $2x-5y=-26$
21. $x=-14$
22. $y=24$
23. yes
24. yes
25. (-4, 37) (0, 9) (3, 9)

Practice Test B

1. D
2. D
3. A
4. A
5. C
6. B
7. C
8. D
9. B
10. D
11. A
12. B
13. B
14. D
15. C
16. B
17. B
18. A
19. C
20. A
21. A
22. B
23. C
24. A

Chapter 4
Section 4.1

1. Independent
3. Consistent
5. Inconsistent
7a. Yes
7b. No
9a. No
9b. Yes
11. (-1, -2)
13. Infinite Solutions
15. Infinite lines; Identical lines
17. 1 solution; Intersecting lines

Section 4.2

1. (-3, 9)
3. All Real Numbers
5. (-2, 2)
7. (-2, -1)
9. Infinite Solutions

11. $\left(-\frac{1}{3}, \frac{2}{3}\right)$
13. No solution
15. No solution
17. $\left(\frac{1}{4}, \frac{1}{5}\right)$
19. $\left(\frac{28}{13}, -\frac{94}{13}\right)$

Section 4.3

1. (2, -3)
3. $\left(-\frac{16}{25}, -\frac{2}{5}\right)$
5. All Real Numbers
7. No Solution
9. All Real Numbers
11. (56, 65)
13. $\left(\frac{10}{17}, \frac{14}{17}\right)$
15. (-1, 3)
17. $\left(\frac{11}{71}, \frac{14}{71}\right)$
19. (1, 4)

Section 4.4

1. $(3,3,2)$
3. $\varnothing$
5. $(3,2,2)$
7. $(1,4,1)$
9. $(3,-4,6)$
11. $(1,-2,4,-3)$

Section 4.5

1. $\{-3,-4\}$
3. adult = $29 and child = $18
5. 4 mph, and 2 mph
7. 150 units
9. After 400 rentals
11. 45°, 50°, & 85°

Practice Test A

1. Yes
2. No
3. $(6,-1)$
4. $(5,-3)$
5. $(-4,-5)$
6. $(1,-1)$
7. $(-3,4,-6)$
8. $\{(x,y)\,|\,10x-12y=6\}$
9. $(-2,3)$
10. $(-1,3,4)$
11. $\varnothing$
12. $\{3,-6\}$
13. 285
14. 54 adult tickets and 28 children tickets
15. 20,000 units

Practice Test B

1. C
2. B
3. B
4. A
5. B
6. D
7. B
8. C
9. A
10. D
11. C
12. D
13. C
14. C
15. B

Chapter 5
Section 5.1

1. Base, exponent
3. Multiply
5. 1
7. 0.000027
9. $2^{11}x^{18}$
11. $x^{19}y^{6}$

13. $(-7)^{48}$
15. $a^{24}b^{30}$
17. $0.0008\,y^{48}$
19. $\dfrac{4x^2z^4}{y^{10}}$
21. 1
23. $\frac{1}{9}$
25. $16mn^3$
27. It will be 64 times larger

Section 5.2

1. Binomial
3. True
5. Coefficient
7. 6th degree; Trinomial
9. 8th degree; None of these
11. 0
13. 4
15. 15ab - 6a - 8b
17. $-10x-13$
19. x^3-7x^2+4x+3
21. $-2m^2+12m-10$
23. 112 ft

Section 5.3

1. $8x^5-6x^3$
3. $-5x^3y+7x^2y^2-3y^4$
5. $\dfrac{1}{10}x^2-\dfrac{23}{10}x+\dfrac{1}{12}$
7. $ab-10a+7b-70$
9. $9x^4+6x^2+1$
11. $x^3-6x^2+10x-4$
13. a^2-3a-4
15. $10x^3+22x^2-x-1$
17. $6x^2+16x+2$

Section 5.4

1. $x^2 - x - 42$
3. $33y^2 - 80xy + 48x^2$
5. $16m^2 + 6mp - p^2$
7. $x^2 - \dfrac{1}{2}x + \dfrac{1}{16}$
9. $49s^2 - 42st + 9t^2$
11. $x^2 - 25$
13. $81x^2 - y^2$
15. $81x^2 - 169$
17. $625x^4 - 1000x^3 + 600x^2 - 160x + 16$

Section 5.5

1. Scientific notation
3. x^7
5. $\dfrac{1}{512}$
7. $\dfrac{-4}{x^6}$
9. $\dfrac{-8}{9}$
11. $\dfrac{1}{x^4}$
13. $-16b^{11}$
15. 1.67×10^{-6}
17. 0.0000003478
19. $200,000$

Section 5.6

1. $-3x^2 + \dfrac{5}{2}x - 2$
3. $-3x^2 + x - \dfrac{4}{x^3}$
5. $x + 1$
7. $4m^2 + 6m - 8 + \dfrac{10}{2m + 3}$
9. $2b^2 + b + 2 - \dfrac{12}{b + 4}$
11. m

Section 5.7

1. $x + 1$
3. $x^2 - 3x + 9$
5. $9x - 18 + \dfrac{28}{x + 2}$
7. $3x^2 + 4x - 8 + \dfrac{20}{x + 1}$
9. -204
11. 122
13. 67
15. $5x^3 - \dfrac{29}{3}x^2 + \dfrac{58}{3}x - \dfrac{113}{3} + \dfrac{223}{3(x + 2)}$

Practice Test A

1. $\dfrac{1}{512}$
2. -64
3. $\dfrac{r^{15}s^{11}}{2}$
4. $-108m^{15}n^7$
5. 1.372×10^{11}
6. 0.00000004284
7. 5
8. $x^2y + 5xy^2 + 2y^2$
9. $9a^2 + 6a - 2$
10. $-10x^2 + 8x + 9$
11. $-2c^2 + 2c - 6$
12. $2a^2 - ab - 3b^2$
13. $-4x^4 + 5x^3 - 2x^2 + 8x$
14. $4m^3 - 16m^2 - 6mn + 2m^2n + n^2$
15. $x^2 - 169$
16. $y^2 - 14y + 49$
17. $49x^2 - \dfrac{1}{4}$
18. $c^3 + 12c^2 + 48c + 64$
19. $2x^2 - 5x$
20. -252 ft
21. $2x^2 - 4x - 3$
22. $\dfrac{3}{4}m - n - \dfrac{2}{n}$
23. $2x + 7$
24. $x^2 + 5x + 25$
25. -23

Practice Test B

1. A
2. D
3. D
4. D
5. B
6. C
7. A
8. A
9. C
10. D
11. B
12. A
13. C
14. D
15. A
16. C
17. B
18. A
19. D
20. D
21. C
22. C
23. D
24. A
25. B

Chapter 6
Section 6.1

1. Factors
3. Greatest common factor
5. 3
7. 1
9. $4y^3$
11. $3a^4b^3$
13. $3(3x^2 - 6x + 2)$
15. $(x^2 + 2)(y + 3)$
17. (b − 4) (a + 5b)
19. 3(8a − 9) (7b + 4)

Section 6.2

1. (x-6) (x+3)
3. (x + 8) (x + 4)
5. (x + 1) (x − 3)
7. (x + 4y)(x − 5y)

9. prime
11. $4y(4x^2 - 4x - 45)$
13. $\frac{1}{2}(x+4)(x-8)$
15. b = 7

Section 6.3

1. (3x + 2)(2x − 1)
3. prime
5. (3x -5)(9x − 4)
7. x(4x + 3) (x − 3)
9. 7(8m − 9)(m + 1)
11. $(4x-5)^2$
13. $4(2x-3)^2$
15. $\left(6x - \frac{1}{2}\right)^2$
17. $\frac{1}{12}(3x-2)(x-2)$

Section 6.4

1. (x + 2)(x + 3)
3. (a − 4)(2a − 5)
4. (2x − 1)(6x − 7)
5. 12(3x + 1)(x − 2)
7. prime
9. $\frac{1}{4}(2x-1)(x-6)$

Section 6.5

1. Difference of squares
3. Perfect square trinomial
5. (x + 7)(x − 7)
7. (11m + 10n)(11m − 10n)
9. prime
11. $\left(\frac{5}{6}t + \frac{2}{3}\right)\left(\frac{5}{6}t - \frac{2}{3}\right)$
13. $(y+3)(y^2 - 3y + 9)$
15. prime
17. $\left(x^2 + y^3\right)\left(x^2 - y^3\right)$
19. $(x-4)\left(x^2 + x + 7\right)$

Martin-Gay *Beginning and Intermediate Algebra, Fourth Edition*   499

Section 6.6

1. $\{-3, 4\}$
3. $\left\{\dfrac{7}{3}, \dfrac{1}{2}\right\}$
5. $\{-7\}$
7. $\{0, 1, -1\}$
9. $\{0, -3, 8\}$
11. $\{0, 3, -1\}$
13. $\left\{\dfrac{1}{2}, -6\right\}$
15. $\{8, 0\}$

Section 6.7

1. 14, 19, 52
3. 12, 14

5. 11, 13
7. L = 8 in, W = 6 in.
9. $\dfrac{9}{4}$ and $\dfrac{1}{4}$
11. 12, 16

Practice Test A

1. $(x - 7)(x + 4)$
2. $(2x + 5)(3x - 2)$
3. $(2x+1)\left(4x^2 - 2x+1\right)$
4. $3(x - 9)(x - 1)$
5. $\left(x^2 +4\right)(x+2)(x-2)$
6. $-(x- 4)(4x - 7)$
7. $(x-2)\left(3x^2 -5\right)$
8. $(2x + 13)(2x - 13)$
9. $(x + 6)(x + 5)$
10. prime
11. $\{4, 9\}$
12. $\{0, -1, -6\}$
13. $\left\{\dfrac{1}{5}, -\dfrac{3}{2}\right\}$
14. $\left\{-\dfrac{2}{5}, -\dfrac{2}{3}\right\}$
15. $\left\{-\dfrac{4}{5}\right\}$

16. $\left\{1, -\dfrac{1}{6}\right\}$
17. $\left\{\dfrac{1}{11}, -2\right\}$
18. $\left\{\dfrac{1}{5}\right\}$
19. $\{0, -5\}$
20. 4.5 seconds

Practice Test B

1. A
2. B
3. C
4. A
5. B
6. D
7. B
8. A
9. A
10. A
11. C
12. C
13. B
14. D
15. B
16. B
17. C
18. D
19. B
20. C

Chapter 7
Section 7.1

1. -1
3. Domain
5. 1
7. $\{x\,|\,x \text{ is a real number, } x \neq -1, x \neq -2\}$
9. $\{x\,|\,x \text{ is a real number, } x \neq -2,\ x \neq 2\}$
11. $\dfrac{x^2}{x-2}$
13. $\dfrac{x-3}{x-2}$
15. $\dfrac{-2y+5}{7-3y}, \dfrac{-(2y-5)}{7-3y}, \dfrac{2y-5}{-(7-3y)}, \dfrac{2y-5}{3y-7}$
17.a. \$200 million
17.b. \$500 million

19. $\dfrac{x+3}{4}$

Section 7.2

1. $\dfrac{5w}{9x^2}$

3. $\dfrac{4}{x+3}$

5. $\dfrac{1}{z+1}$

7. $\dfrac{7y^2}{2x^2}$

9. $\dfrac{x-4}{x-3}$

11. $\dfrac{(x-3)(x-1)}{x+1}$

13. 1

15. $\dfrac{(x+3)^2(x-1)}{(x+4)^2(x-2)}$

Section 7.3

1. $\dfrac{3x-10}{x+2}$

3. $\dfrac{-6x-4}{x+3}$

5. $\dfrac{-5x+8}{x+1}$

7. $(x+1)(x-3)$
9. $(x+4)(x-4)(x-4)$
11. $(x+1)(x-1)(x^2-x+1)$

13. $\dfrac{3x(x-3)}{x^3-5x^2+6x}$

15. $\dfrac{(x-1)(x+2)}{x^3+6x^2-x-6}$

Section 7.4

1. $\dfrac{7x-4}{x^2-1}$

3. $\dfrac{-y-4}{y+3}$

5. $\dfrac{2\left(x^2+7x+6\right)}{(x+3)(x+3)(x+5)}$

7. $\dfrac{-2(y+1)}{(y-4)(y+3)(y+1)}$

9. $\dfrac{2x^2+3x-1}{(x-8)(x-1)(x+1)}$

11. $\dfrac{2(x^2-x-23)}{(x+1)(x-5)(x-6)}$

13. $\dfrac{x^2+15x-22}{(x-6)(x-2)(x+2)}$

Section 7.5

1. $\left\{\dfrac{7}{5}\right\}$

3. $\{8, -2\}$
5. No solution
7. Mo solution

9. $\left\{-\dfrac{9}{13}\right\}$

11. $p=\dfrac{qr}{q-r}$

13. $B=\dfrac{2U+TE}{T}$

15. Answers may vary

Section 7.6

1. 33
2. – 3
3. 7
4. -4
5. 60 women
6. 915 dentists
7. 4560 m
8. $\dfrac{11}{3}$
9. 4
10. 4 and 6
11. 3.5 minutes
12. 2 hours
13. 3.75 hours
14. motorcycle is 60 mph and the car is 70 mph

15. Lucy walks at 1 mph, Michelle walks at 2 mph

16. 150 mph

Section 7.7

1. $\dfrac{3}{2}$

3. $2x + y$

5. $\dfrac{3y - 4}{2y + 2}$

7. $\dfrac{2 - 4xy}{5y + 3x}$

9. $\dfrac{2b^2 + 3a}{b(b - a)}$

11. $\dfrac{x\left(45x^2 - 1\right)}{3x^3 + 1}$

Practice Test A

1. $\{3, 2\}$
2. $\{2, -2\}$
3. $\{0, 6\}$

4. $\dfrac{x^2 + 2x - 7}{x + 1}$

5. $\dfrac{2x}{(x + 4)(x - 5)}$

6. $\dfrac{(x - 2)^2}{(x + 2)(2x + 1)}$

7. $\dfrac{2x}{(x - 5)(x + 3)}$

8. -1

9. $\dfrac{(x - 4)(x + 1)}{3x}$

10. $\dfrac{5}{8(x + 2)}$

11. $\dfrac{x(2x - 1)}{4 - 3x^2}$

12. $\dfrac{4x^2 - x + 3}{5x + 9}$

13. $-\dfrac{3}{7}$

14. $\{3, 2\}$

15. 8

16. $\dfrac{m}{1 - mt}$

17. $B = \dfrac{bh - 2A}{-h}$

18. 1.2 hours

19. 15

Practice Test B

1. B
2. A
3. C
4. D
5. A
6. B
7. A
8. C
9. B
10. C
11. C
12. D
13. D
14. C
15. B
16. D
17. D
18. A
19. B

Chapter 8
Section 8.1

1. Horizontal
3. Vertical
5.

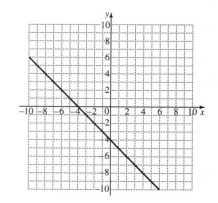

7.

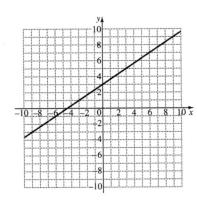

9. $f(x) = \dfrac{1}{4}x - 5$

11. $f(x) = -7x + \dfrac{41}{10}$

13. $f(x) = \dfrac{1}{3}x - \dfrac{10}{3}$

15. $f(x) = -\dfrac{2}{3}x - 23$

17. slope = 0

19. $f(x) = -\dfrac{13}{19}x + \dfrac{43}{380}$

Section 8.2

1. $y = |x|$
3. $f(9) = 3.6$
5. -4
7. 5
9. $-\dfrac{4}{5}$
11. not a real number
13.

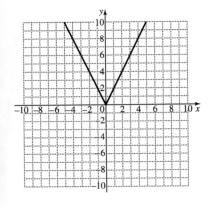

15.

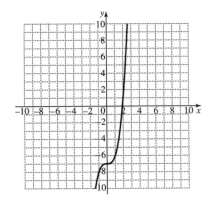

17. The second graph shifts 2 units to the right, and 5 units up.
19. 3

Section 8.3

1.

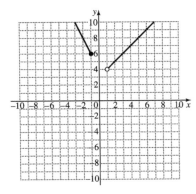

3.

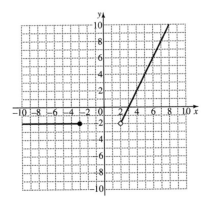

5.

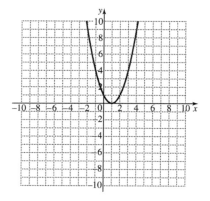

11.

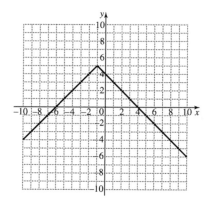

7.

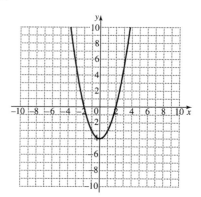

13.

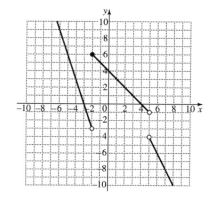

9.

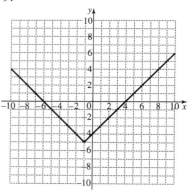

Section 8.4

1. $k = 7;\ y = 7x$

3. $k = \dfrac{4}{5};\ y = \dfrac{4}{5}x$

5. $k = \dfrac{9}{8};\ y = \dfrac{9}{8}x$

7. $k = 24;\ y = \dfrac{24}{x}$

9. $k = 4;\ y = \dfrac{4}{x}$

11. 54 mph

13. $k = \dfrac{3}{5};\ y = \dfrac{3}{5}x^2 z$

15. 22.5 tons

17. y is one-fourth its original amount.

Martin-Gay *Beginning and Intermediate Algebra, Fourth Edition*

Practice Test A

1. 9
2. 1, 3
3.

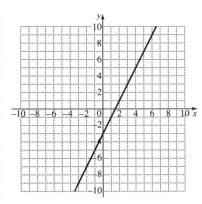

4.

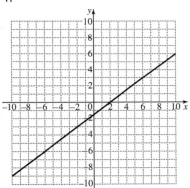

5.

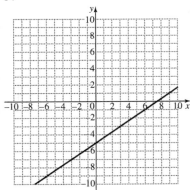

6.

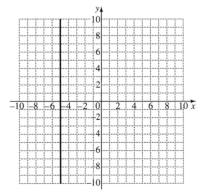

7.

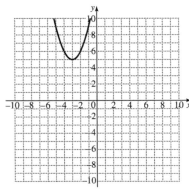

8.

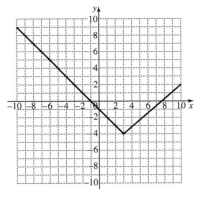

9.

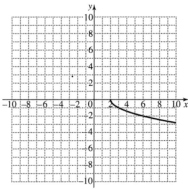

10. $f(x) = -3x + 16$

11. $f(x) = \dfrac{1}{3}$

12. $f(x) = \dfrac{12}{5}x + \dfrac{4}{5}$

13. $f(x) = \dfrac{1}{3}x - \dfrac{11}{3}$

14. parallel

15.

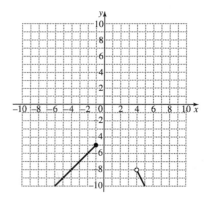

16.

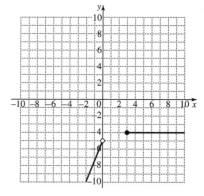

17. $k = 6, \; y = 6x^2$

Practice Test B

1. B
2. A
3. A
4. B
5. B
6. C
7. D
8. B
9. A
10. D
11. A
12. B
13. B
14. A
15. B
16. B
17. D

Chapter 9
Section 9.1

1. Compound
3. Union; $\cup$
5. $\{1, 15, 19, 23, 29\}$
7. $\varnothing$
9. $[-4, 8]$
11. $[3, 10]$
13. $\left[-3, \dfrac{3}{2}\right]$
15. $\{x \,|\, x \text{ is an odd number or } x = 2 \text{ or } x = 12\}$
17. $(-\infty, 2)$
19. $(-4, \infty)$
21. $(-\infty, 1) \cup (5, \infty)$
23. $\left(-\dfrac{4}{7}, \dfrac{10}{3}\right]$

Section 9.2

1. 8, -8
3. 10, -10
5. 3, -2

Martin-Gay *Beginning and Intermediate Algebra, Fourth Edition*

7. 8, 4
9. 2
11. ∅
13. $|x+5|=4$

Section 9.3

1. $(-5,5)$
3. $(-3,9)$
5. $[-16,4]$
7. ∅
9. $[-1,8]$
11. $(-\infty,-4]\cup[6,\infty)$
13. $(-\infty,-21)\cup(27,\infty)$
15. $(-\infty,-2]\cup[8,\infty)$

Section 9.4

1. Boundary Line
3. False
5. Solution

7.

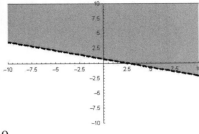

9.

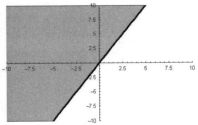

11.

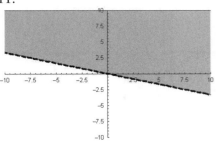

13.

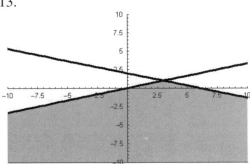

15.

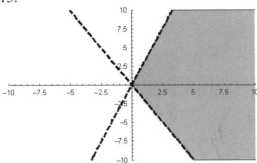

17.

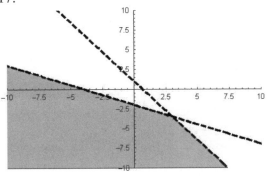

19.

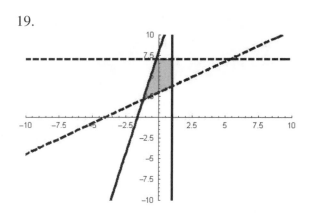

Practice Test A

1. 7, - 3
2. ∅
3. $\frac{7}{5}, -9$
4. 6
5. $(2,8)$
6. $(-2,5)$
7. $(-\infty, -13) \cup (-3, \infty)$
8. $(-\infty, -5) \cup [2, \infty)$
9. $\left(\frac{5}{2}, \frac{9}{2}\right]$
10. $[-1, 5)$
11. $\{16, 18, 20, 22, 24\}$
12.

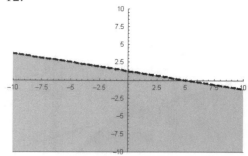

13.

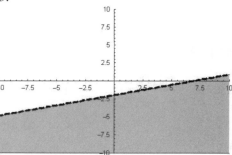

14.

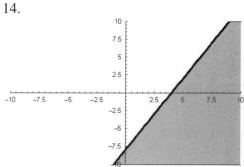

15.

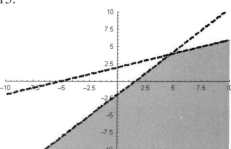

16.

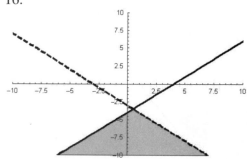

Practice Test B

1. D
2. A
3. B
4. A
5. B
6. D
7. B
8. A

Martin-Gay *Beginning and Intermediate Algebra, Fourth Edition*

9. C
10. B
11. D
12. C
13. A
14. D
15. B
16. C

Chapter 10
Section 10.1

1. index; radicand; radical sign
3. squares
5. 60
7. 0.5
9. Not a real number
10. $9x^8$
12. 10.954
14. 16.125
16. $-\dfrac{1}{3}$
18. 10
20. $2x^6$
22. $x-5$
24. 5

26.

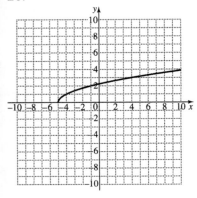

28.

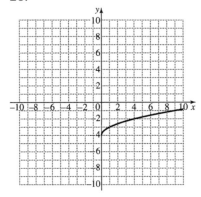

30. 50

Section 10.2

1. 5
3. $\sqrt[3]{5x}$
5. 216
7. 16
9. $\dfrac{1}{2}$
11. $\dfrac{4x^{3/4}}{5}$
13. $x^{3/4}$
15. $5^{7/12}$
17. $x^{5/6}+x^{1/2}-2x^{1/3}-2$
19. $x^{3/5}\left(x^{1/5}-3x^{2/5}\right)$
21. $\sqrt{y}$
23. $\sqrt{x^7 y^5}$
25. $\sqrt[6]{4500}$
27. $x^{-5/12}=\dfrac{1}{x^{5/12}}$

Section 10.3

1. $\sqrt{35}$
3. $\sqrt{\dfrac{55}{6}}$
5. $\sqrt[3]{\dfrac{3}{5}}$

7. $\dfrac{9}{x\sqrt{z}}$

9. $\dfrac{z^8}{5x^6}$

11. $4x\sqrt{2}$

13. $-2x^2\sqrt[5]{y}$

15. 5; 5

17. $2\sqrt{26}$; 10.198

19. $\left(\dfrac{3}{2}, -1\right)$

21. $\left(-\sqrt{3}, \dfrac{7}{2}\sqrt{5}\right)$; $\sqrt{173}$

Section 10.4

1. $17\sqrt{2} - 15\sqrt{5}$

3. $\left(y^2 + 3y + 1\right)\sqrt{y}$

5. $\dfrac{-2\sqrt{3x}}{5}$

7. $6\sqrt{2} - 12$

9. $-16 - 3\sqrt{5}$

11. $7 - 4\sqrt{3}$

13. $22\sqrt{3}$ feet

Section 10.5

1. Rationalizing the numerator

3. $\dfrac{\sqrt{5}}{\sqrt{5}}$

5. $\dfrac{\sqrt{30x}}{6x}$

7. $4\sqrt[5]{4x}$

9. $\dfrac{\sqrt{10} + 4\sqrt{5} + \sqrt{6} + 4\sqrt{3}}{14}$

11. $\dfrac{7\left(3 + \sqrt{x}\right)}{9 - x}$

13. $\dfrac{x}{5\sqrt{3x}}$

15. $\dfrac{1 - x}{3\left(1 - \sqrt{x}\right)}$

17. $\dfrac{2 - 16x^2}{-4x^2 + 2 - 3x\sqrt{2}}$

Section 10.6

1. 35
3. $\varnothing$
5. -12
7. 0
9. 1
11. 202 miles
13. 1 and 0

Section 10.7

1. Imaginary Unit
3. Pure imaginary
5. Real
7. $6i$
9. $i\sqrt{5}$
11. $7 - i$
13. $4 + 9i$
15. 60
17. $-15 - 23i$
19. 20

21. $\dfrac{8}{5} + \dfrac{4}{5}i$

23. $4 + i$
25. -1
27. $64i$

29. $\dfrac{1}{2} - \dfrac{3\sqrt{2}}{7}i$

Practice Test A

1. $8\sqrt{5}$

2. $2\sqrt[5]{x^4}$

3. $\dfrac{3}{2}$

4. D $x^{7/6} - x^{3/4}$

5. $a^6 b^9$

6. $\dfrac{7x^{1/3}}{y^{2/5}}$

7. $3a\sqrt{3}i$

8. $8i$

9. $\dfrac{4\sqrt{5}}{5}$

10. $\dfrac{x+8\sqrt{x}+15}{9-x}$

11. $\dfrac{\sqrt[3]{x^2}}{x}$

12. $\dfrac{21}{37}+\dfrac{22}{37}i$

13. $30-4\sqrt{2}$

14. $\sqrt{15}-\sqrt{10}$

15. $7-2x\sqrt{7}+x^2$

16. $13x\sqrt{2}+2\sqrt{2x}$

17. 13. 115

18. 4.611

19. 22

20. 6 and -1

21. 4

22. $27+11i$

23. $1-3i$

24. $\sqrt{146}$

25. $\left(\frac{5}{2},2\right)$

Practice Test B

1. A
2. B
3. D
4. B
5. C
6. A
7. A
8. B
9. A
10. D
11. A
12. C
13. B
14. B
15. A
16. D
17. A
18. C
19. A
20. C
21. A
22. D

23. B
24. C
25. D

Chapter 11
Section 11.1

1. Quadratic Equation
3. Completing the Square
5. 6 and -6
7. $\pm\sqrt{5}$
9. $3\pm 3i$
11. $-3\pm\sqrt{7}$
13. $\dfrac{-3\pm\sqrt{29}}{2}$
15. $-3\pm\sqrt{3}i$
17. $-\dfrac{3}{2}\pm\dfrac{\sqrt{13}i}{2}$
18. 20%
19. 40%
20. $24\sqrt{2}$
21. $6\sqrt{2}$

Section 11.2

1. -15 and -1
3. $\dfrac{-3\pm\sqrt{57}}{4}$
5. $-\dfrac{1}{4}\pm\dfrac{\sqrt{39}}{4}i$
7. Two real solutions
9. Two complex solutions
11. 14 ft
13. $\dfrac{-\sqrt{5}\pm\sqrt{17}}{2}$

Section 11.3

1. 5 and 1
3. 3
5. 1 and $-\frac{1}{2}$
7. $-\frac{20}{3}$
9. $\pm 1, \pm 2i$
11. $7-2\sqrt{6}$

13. 5 mph, then 4mph

15. Joe will take 15.1 hours, and
Sarah will take 17.1 hours

17. 1 and $-\dfrac{1}{5}$

Section 11.4

1. $[-4,3]$

3. $(-6,2)$

5. $(-\infty,-5)\cup(-1,1)\cup(5,\infty)$

7. $(-\infty-3)\cup(-1,\infty)$

9. $(-7,-2)\cup(-2,\infty)$

11. Answers will vary.

Section 11.5

1. axis of symmetry at $x = 0$ Vertex $(0,2)$

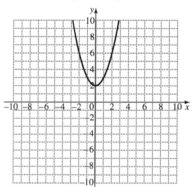

3. axis of symmetry at $x = 0$ Vertex $(0,6)$

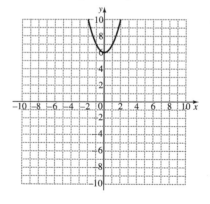

5. axis of symmetry at $x = 4$ Vertex $(4,0)$

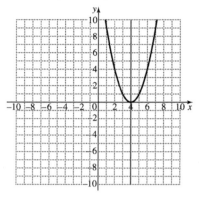

7. axis of symmetry at $x = 2$ Vertex $(2,5)$

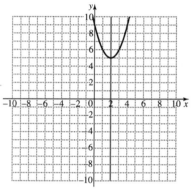

9. axis of symmetry at $x = 3$ Vertex $(3,-2)$

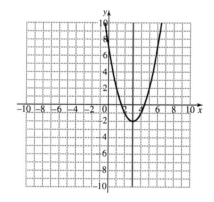

Martin-Gay *Beginning and Intermediate Algebra, Fourth Edition*

11. axis of symmetry at $x = 0$ Vertex $(0,0)$

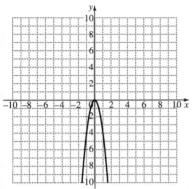

13. axis of symmetry at $x = 0$ Vertex $(0,0)$

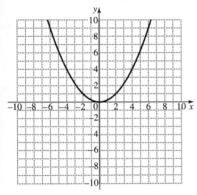

15. axis of symmetry at $x = 1$ Vertex $(1,2)$

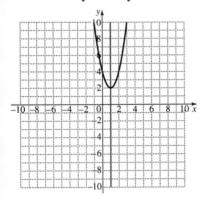

17. $f(x) = -2(x-4)^2 - 6$

Section 11.6

1. $f(x) = (x-1)^2 - 16$; $(1,-16)$

3. $f(x) = \left(x + \frac{5}{2}\right)^2 + \frac{47}{4}$; $\left(-\frac{5}{2}, \frac{47}{4}\right)$

5. $f(x) = -3(x-2)^2 + 20$; $(2,20)$

7. $-\dfrac{b}{2a}$

9. $(-2,3)$

11. $(1,-16)$

13. 144 ft

15. -1.29; minimum

Practice Test A

1. $\dfrac{3 \pm \sqrt{29}}{2}$

2. $\frac{2}{7}$ and $-\frac{8}{5}$

3. $-3 \pm \sqrt{6}$

4. 13 and 1

5. ± 3 and $\pm i$

6. 9 and 11

7. 3, - 3, and - 2

8. $3 \pm 2\sqrt{5}$

9. $6 \pm \sqrt{15}$

10. $\dfrac{6 \pm \sqrt{21}}{2}$

11. $\left(-\infty, \frac{5}{3}\right] \cup [8,\infty)$

12. $\left(-7, \frac{9}{2}\right)$

13. $\left(-\infty, -\frac{3}{2}\right) \cup (1,\infty)$

14. Vertex $(-3,-6)$

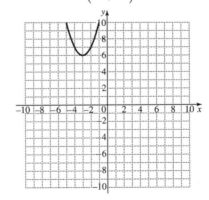

15. Vertex $(0, -4)$

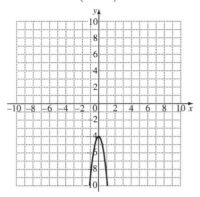

16. Vertex $(-2, 5)$

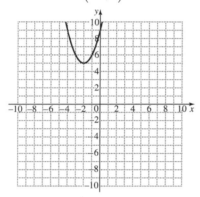

17. Vertex $(4, 7)$

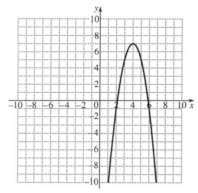

18. 13 and 21
19. height is 4 ft, base is 16 ft.
20. 4.5 seconds

Practice Test B

1. A
2. D
3. B
4. A

5. A
6. C
7. D
8. B
9. C
10. A
11. A
12. C
13. D
14. C
15. B
16. D
17. A
18. B
19. C
20. A

Chapter 12
Section 12.1

1a. $3x - 2$, b. $-x - 8$

c. $2x^2 - 7x - 15$ d. $\dfrac{x-5}{2x+3}$

3a. $\sqrt{x+5} + 4x^2$ b. $\sqrt{x+5} - 4x^2$

c. $4x^2\sqrt{x+5}$ d. $\dfrac{\sqrt{x+5}}{4x^2}$

5. 14
7. 23
9. $\sqrt{x+3} + 7$; $\sqrt{x-4}$
11. $(2x+5)^{\frac{2}{3}} - 9$; $\sqrt[3]{2x-13}$
13. 6
15. 18
17. -6

Section 12.2

1. $y = x$
3. Horizontal
5. x
7. No
9. No
11. $4x - 3$
13. $x^3 + 5$

15. $f^{-1}(x) = \dfrac{x-1}{5}$

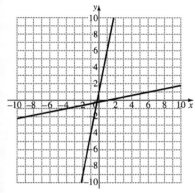

17a. $(-5, 25)$ b. $(25, -5)$

Section 12.3

1.

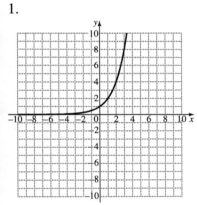

3.

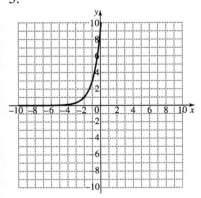

5. -3 7. $-\dfrac{1}{3}$

9. $3429.36

11.

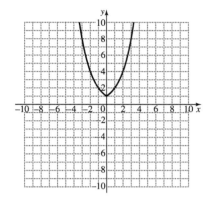

Section 12.4

1. $4^2 = 16$

3. $7^9 = x$

5. $\log_5 \dfrac{1}{125} = -3$

7. $\log_3 81 = 4$

9. 5

11. -2

13. -3

15. 8

17. 3

19.

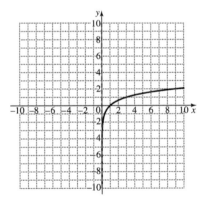

21.

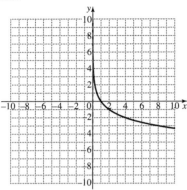

Section 12.5

1. $\log_3 5x$

3. $\log_7 \left(x^5 - 2x^4 \right)$

5. $\log_3 \dfrac{x}{6}$

7. $\log_2 \dfrac{x^2 - 6}{x + 5}$

9. $6 \log_4 x$

11. $\dfrac{1}{2} \log_5 y$

13. $\log_4 4 = 1$

15. $\log_8 4x$

17. $\log_3 5 + 2 \log_3 x$

19. $\dfrac{1}{2} \log_7 x - 2 \log_7 (x + 4)$

21. $b = 3$

Section 12.6

1. 0.4082
3. 3
5. -4
7. $\dfrac{10^{1.1}}{2}$; 6.2946
9. 1.6094
11. 0.02
13. -5
15. 1.5850
17. -2.1053

Section 12.7

1. $\dfrac{\ln 9}{\ln 5}$; 1.3652

3. $\dfrac{\ln 24}{3 \cdot \ln 6}$; 0.5912

5. 4

7. $-\dfrac{7}{8}$

9. $\dfrac{1}{4}$

11. 14 years
13. $x \le 5$

Practice Test A

1. $x - 4$
2. $-6x^2 + 8x$
3. $-6x - 4$
4. -16
5. $2x + 14$
6. $\sqrt[3]{x + 4}$
7. $\dfrac{7x - 8}{5}$

8. $\log_6 \dfrac{3x^2}{x + 2}$

9. $\log_3 \dfrac{5\sqrt{x}}{8}$

10. 1.5440
11. 1024
12. $\dfrac{1}{5}$
13. 4 and -2
14. $\dfrac{23}{8}$
15. 5.1828
16.

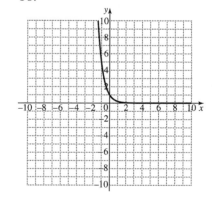

Martin-Gay *Beginning and Intermediate Algebra, Fourth Edition*

17.

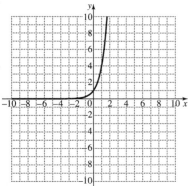

18. $10810.59
19. $14366.61
20. 0.375 grams

Practice Test B

1. B
2. C
3. A
4. A
5. D
6. C
7. B
8. C
9. A
10. B
11. C
12. A
13. D
14. B
15. A
16. C
17. B
18. D
19. C
20. A

Chapter 13
Section 13.1

1. Radius
3. Conic sections
5. Circle; center

7. V: $(-1,-4)$

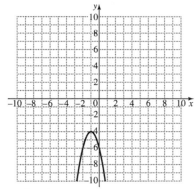

9. V: $(3,3)$

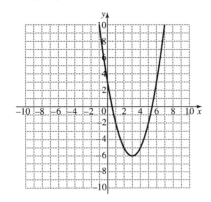

11. C: $(0,3)$; $r = 2$

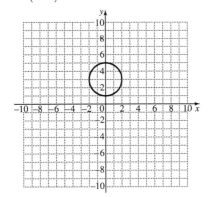

13. C: $(-2,-1)$; r = 4

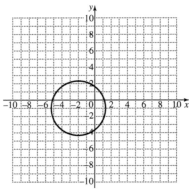

15. $(x+1)^2 + (y+6)^2 = 49$

17. $C:(0,0) \quad r = 4$

19. $C:(3,5) \quad r = \sqrt{38}$

20. Answers will vary

21. If $a < 0$ and x-coordinate in vertex is negative and if $a > 0$ and the x-coordinate in the vertex is positive.

Section 13.2

1. Foci

3. Hyperbola; $(0,a)$ and $(0,-a)$

5.

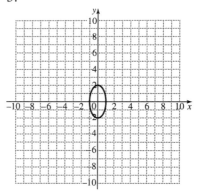

7.

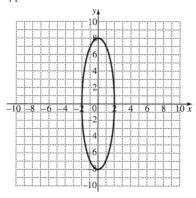

9.

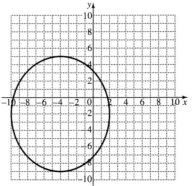

11.

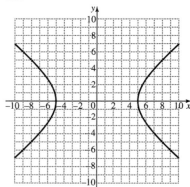

13.

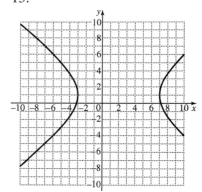

15.

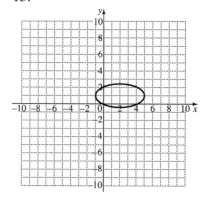

518   Martin-Gay *Beginning and Intermediate Algebra, Fourth Edition*

Section 13.3

1. $(0,-2)(4,0)$

3. $\varnothing$

5. $\left(\frac{12}{5}, -\frac{8}{5} \right)(4,0)$

7. $\varnothing$

9. $(0,-2)\left(\frac{-4\sqrt{5}}{3}, \frac{4}{3} \right)\left(\frac{4\sqrt{5}}{3}, \frac{4}{3} \right)$

11. $-\frac{18}{5}$ and $\frac{24}{5}$ or 6 and 0

Section 13.4

1.

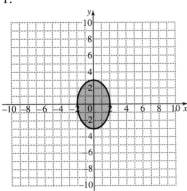

3.

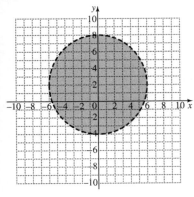

5.

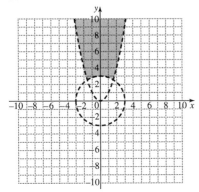

7.

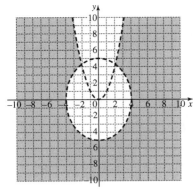

9.

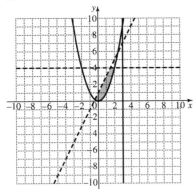

Practice Test A

1.

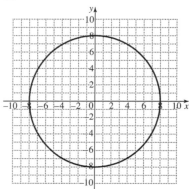

2.

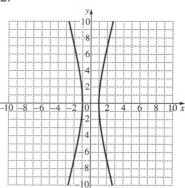

3.

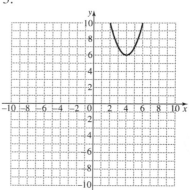

4.

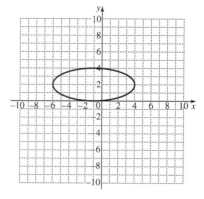

5.

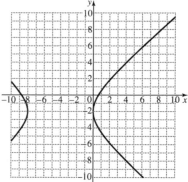

6. $\varnothing$

7. $(0,-2)(2,0)$

8. $\varnothing$

9.

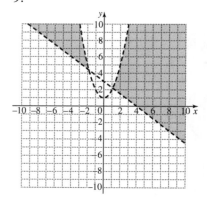

10.

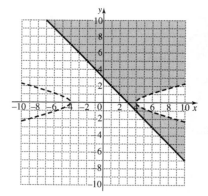

Martin-Gay *Beginning and Intermediate Algebra, Fourth Edition*

11.

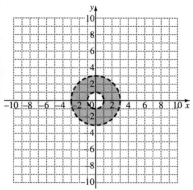

12. 11.1 feet

Practice Test B

1. A
2. B
3. C
4. A
5. D
6. B
7. C
8. D
9. A
10. D
11. B
12. A

Chapter 14
Section 14.1

1. Finite
3. Infinite
5. $1, -3, 9, -27, 81$
7. $\dfrac{1}{4}, \dfrac{1}{5}, \dfrac{1}{6}, \dfrac{1}{7}\dfrac{1}{8}$
9. -18
11. $\dfrac{-1}{400}$
13. 8388608
15. $a_n = -2^n$
17. $a_n = \dfrac{3}{n+4}$
19. $120, $160, $200, $240, $280
21. $1, 1.0125, 1.0167, 1.0189, 1.0202$

Section 14.2

1. Arithmetic; Difference
3. First
5. $25, 32, 39, 46, 53$
7. $10, 26, 42, 58, 74$
9. 34
11. -28
13. $48, 24, 12, 6, 3$
15. $-1, -4, -16, -64, 256$
17. -1458
19. $-\dfrac{1}{32}$
21. $32.3, -71.06, 156.332, -343.9304$

Section 14.3

1. 60
3. 250
5. 496
7. $\displaystyle\sum_{i=2}^{6} 2i + 1$
9. -13
11. 28
13. $\displaystyle\sum_{i=1}^{5} i$

Section 14.4

1. 63
3. 42
5. 484
7. $\dfrac{85}{64}$
9. 64
11. $\dfrac{32}{9}$
13. $\dfrac{4}{9}$

Section 14.5

1. $m^3 + 3m^2 n + 3mn^2 + n^3$
3.
$$x^8 + 8x^7 y + 28x^6 y^2 + 56x^5 y^3$$
$$+ 70x^4 y^4 + 56x^3 y^5 + 28x^2 y^6 +$$
$$8xy^7 + y^8$$
5. 255024

7. $x^5 - 5x^4y + 10x^3y^2 - 10x^2y^3 + 5xy^4 - y^5$

9. $625a^4 - 1000a^3b + 600a^2b^2 - 160ab^3 + 16b^4$

11. $700000x^4z^4$

13. $189m^5n^2$

15. $840x^3$

Practice Test A

1. $\dfrac{3}{5}, \dfrac{5}{6}, 1, \dfrac{9}{8}, \dfrac{11}{9}$

2. $\dfrac{5}{6561}$

3. $a_n = 7\left(-\dfrac{1}{7}\right)^{n-1}$

4. $a_n = 24 + (n-1)(-14)$

5. 161

6. $-21\big/8$

7. $864\big/5$

8. -104

9. 88

10. 3276

11. $16x^4 - 32x^3y + 24x^2y^2 - 8xy^3 + y^4$

12.
$a^5 + 15a^4b + 90a^3b^2 + 270a^2b^3 + 405ab^4 + 243b^5$

Practice Test B

1. D
2. A
3. B
4. A
5. C
6. B
7. D
8. B
9. C
10. A
11. D
12. B